Discovering Hidden Worlds

The Fascinating Stories
of Pioneering Explorers

Cyril Daniels

The presentation of the information is without
contract or any type of guarantee assurance. The
trademarks that are used are without any consent,
and the publication of the trademark is without
permission or backing by the trademark owner. All
trademarks and brands within this book are for
clarifying purposes only and are the owned by the
owners themselves, not affiliated with this document.

Table of Contents

Chapter 1

Introduction

The Allure of Exploration

From the earliest whispers of human history, the desire to explore has been a defining characteristic of our species. This innate curiosity, a compelling force driving individuals to venture beyond the confines of their known world, has led to some of the most significant discoveries and advancements in human civilization. The allure of exploration is not merely about the physical act of traversing new lands or seas; it is deeply rooted in a quest for knowledge, adventure, and the unknown.

The fascination with what lies beyond the horizon has been a constant theme throughout history. Ancient civilizations, such as the Egyptians, the Phoenicians, and the Greeks, embarked on voyages across the Mediterranean and into the unknown, seeking new trade routes, resources, and knowledge. These early explorers laid the groundwork for future generations, inspiring a legacy of exploration that would span centuries.

One of the most compelling aspects of exploration is the promise of discovery. The unknown holds an almost magical allure, a promise of new lands, cultures, and opportunities. For many explorers, the drive to uncover what lies beyond their immediate surroundings is a powerful motivator. This curiosity is not limited to professional explorers; it is a universal human trait. Whether it is a child exploring their backyard, an astronaut venturing into space, or a scientist delving into the mysteries of the human genome, the desire to discover something new is a fundamental part of the human experience.

The impact of discoveries on society cannot be overstated. Each new discovery has the potential to reshape our understanding of the world and our place in it. The voyages of Christopher Columbus, for example, opened up the Americas to European exploration and colonization, leading to profound changes in global trade, culture, and politics. The discovery of the New World also had significant consequences for the indigenous populations, leading to both cultural exchanges and devastating conflicts.

Technological advancements have played a crucial role in enabling exploration. The development of new tools and techniques has allowed explorers to venture further and more safely than ever before. The invention of the compass, for example, revolutionized navigation, making it possible for sailors to travel long distances with greater accuracy and confidence. Similarly, the development of more advanced ships, such as the caravel, enabled explorers to undertake longer and more ambitious voyages.

The evolution of exploration has been marked by a series of key milestones and turning points. The Age of Discovery, spanning the 15th to the 17th centuries, saw European explorers such as Vasco da Gama, Ferdinand Magellan, and John Cabot chart new territories and establish trade routes that would shape the course of history. This period was characterized by a spirit of adventure and a desire to expand the boundaries of the known world.

In the 18th and 19th centuries, exploration took on new dimensions with the advent of scientific expeditions. Figures such as Captain James Cook and Charles Darwin embarked on journeys that not only mapped new territories but also contributed to our understanding of the natural world. These explorers combined their adventurous spirit with a rigorous scientific approach, collecting data and specimens that would form the basis of modern biology, geology, and anthropology.

The allure of exploration is also deeply connected to the human spirit of perseverance and resilience. The challenges faced by explorers, from harsh climates to unknown dangers, require a unique combination of courage, determination, and resourcefulness. The stories of explorers such as Ernest Shackleton, who led his crew to safety after their ship was trapped in Antarctic ice, are testaments to the extraordinary human capacity to overcome adversity.

Exploration is not without its ethical and moral considerations. The impact of exploration on indigenous populations, ecosystems, and cultures has been profound and often devastating. The drive to discover new lands and resources has sometimes led to exploitation, displacement, and environmental degradation. As we continue to explore new frontiers, it is essential to balance the desire for discovery with a respect for the rights and well-being of those who inhabit these regions.

The allure of exploration continues to captivate the human imagination in the modern era. Space exploration, for example, represents the next frontier in our quest to understand the universe. Missions to the Moon, Mars, and beyond are driven by the same curiosity and desire for discovery that motivated early explorers. The potential for new discoveries, from the possibility of extraterrestrial life to the understanding of our own planet's origins, fuels our fascination with space exploration.

In addition to space, the depths of our oceans remain one of the least explored regions on Earth. Advances in technology, such as remotely operated vehicles and submersibles, have allowed scientists to explore the deep sea and uncover its many mysteries. The discovery of hydrothermal vents, deep-sea ecosystems, and new species highlights the incredible biodiversity that exists in these remote environments.

The quest for exploration is also evident in the field of archaeology, where researchers continue to uncover the remnants of ancient civilizations and gain insights into human history. The discovery of the tomb of Tutankhamun, the ruins of Pompeii, and the ancient city of Machu Picchu are just a few examples of how archaeological exploration has enriched our understanding of the past.

Ultimately, the allure of exploration is a testament to the boundless curiosity and adventurous spirit of humanity. It is a reflection of our desire to push the boundaries of what is known and to seek out new experiences and knowledge. This drive has led to some of the most remarkable achievements in human history and continues to inspire us to explore new frontiers.

As we look to the future, the possibilities for exploration are limitless. Advances in technology, science, and our understanding of the world will undoubtedly open up new avenues for discovery. Whether it is the exploration of distant planets, the depths of the oceans, or the mysteries of the human mind, the allure of exploration will continue to drive us forward, shaping the course of human history and expanding our horizons.

The Impact of Discoveries on Society

Throughout history, discoveries have acted as catalysts for change, reshaping societies, economies, and the very fabric of human existence. Each significant discovery, from new lands to groundbreaking scientific theories, has reverberated through time, leaving a legacy that is often both profound and far-reaching. The impact of these discoveries on society is a rich tapestry woven with threads of innovation, conflict, and transformation.

The Age of Discovery, beginning in the 15th century, brought about some of the most dramatic changes in human history. European explorers, driven by the pursuit of new trade routes and resources, embarked on voyages that expanded the known world. The discovery of the Americas by Christopher Columbus in 1492 is perhaps the most iconic event of this era. This single event set off a chain reaction of exploration, colonization, and cultural exchange that fundamentally altered the course of global history.

The immediate impact of such discoveries was often economic. The influx of precious metals from the New World, particularly silver from mines in Mexico and Peru, had a profound effect on European economies. It led to the Price Revolution, a period of inflation that transformed economic practices and the social order. The newfound wealth fueled the growth of trade and the rise of mercantilism, laying the groundwork for the modern capitalist economy.

However, the impact of these discoveries extended far beyond economics. They also initiated a complex web of cultural exchanges and conflicts. The arrival of Europeans in the Americas brought about the Columbian Exchange, a widespread transfer of plants, animals, and diseases. This exchange had a profound impact on societies on both sides of the Atlantic. European diets were enriched by crops such as potatoes, maize, and tomatoes, which became staples in various cuisines. Conversely, Old World diseases like smallpox and measles decimated indigenous populations in the New World, leading to dramatic demographic shifts and the collapse of several advanced civilizations.

The impact on indigenous cultures was profound and often tragic. The arrival of European settlers and explorers led to the displacement and marginalization of native populations. In many cases, entire cultures were eradicated or irrevocably altered. The Spanish conquest of the Aztec and Inca empires, for example, resulted in the destruction of complex societies and the imposition of European culture and religion.

These events underscore the often harsh realities of exploration and discovery, highlighting the dual-edged nature of progress.

The scientific discoveries of the Renaissance and Enlightenment periods also had a transformative impact on society. Figures such as Galileo Galilei, Isaac Newton, and Nicolaus Copernicus challenged existing paradigms and laid the foundations for modern science. Their discoveries revolutionized our understanding of the natural world and our place within it. The heliocentric model of the solar system, proposed by Copernicus and later confirmed by Galileo, fundamentally altered humanity's view of the universe. Newton's laws of motion and universal gravitation provided a framework that has guided scientific inquiry for centuries.

These scientific advancements had far-reaching implications for society. They contributed to the decline of religious dogma as the dominant explanation for natural phenomena, paving the way for a more secular and empirical approach to understanding the world. The Enlightenment, fueled by these scientific discoveries, emphasized reason, individualism, and skepticism of authority. This intellectual movement laid the groundwork for modern democratic societies, influencing political thought and the development of institutions that prioritize human rights and freedoms.

The Industrial Revolution, another period of profound discovery and innovation, transformed societies in unprecedented ways. The development of new technologies, such as the steam engine, mechanized textile production, and the telegraph, revolutionized industries and economies. These advancements led to the mass production of goods, increased efficiency, and the rise of factory-based economies. The Industrial Revolution also brought about significant social changes, including urbanization, the rise of a working class, and shifts in family structures and gender roles.

The impact of these technological discoveries was not without its challenges. The rapid pace of industrialization led to harsh working conditions, environmental degradation, and social upheaval. The exploitation of labor, including child labor, and the widening gap between the wealthy and the working poor highlighted the need for social reforms. These issues spurred movements for workers' rights and the establishment of labor unions, which sought to improve conditions and advocate for fair wages and hours.

In the modern era, discoveries in science and technology continue to shape society in profound ways. The advent of the internet and digital technologies has revolutionized communication, commerce, and entertainment. The ability to connect with people across the globe instantaneously has transformed social interactions and created a more interconnected world. The rise of social media platforms has given individuals a voice and a means to organize and advocate for change, but it has also raised concerns about privacy, misinformation, and the impact on mental health.

Medical discoveries have also had a significant impact on society. The development of vaccines, antibiotics, and advanced medical technologies has improved public health and increased life expectancy. These advancements have eradicated or controlled many infectious diseases, contributing to population growth and improved quality of life. However, they also present ethical dilemmas and challenges, such as ensuring equitable access to healthcare and addressing the implications of genetic engineering and biotechnology.

The exploration of space, driven by both governmental and private endeavors, represents the next frontier of discovery. The achievements of space exploration, from landing on the Moon to the deployment of telescopes that peer into the far reaches of the universe, have expanded our understanding of the cosmos and our place within it. These discoveries inspire a sense of wonder and possibility, fueling aspirations for future exploration and the potential colonization of other planets.

The impact of discoveries on society is a testament to the power of human curiosity and ingenuity. Each new discovery, whether it is a new land, a scientific breakthrough, or a technological innovation, carries the potential to transform the world in unexpected ways. The process of discovery is often fraught with challenges and ethical considerations, but it is also a driving force behind progress and the advancement of human knowledge.

The Evolution of Exploration

From the dawn of human civilization, the spirit of exploration has been a driving force behind progress and transformation. This innate desire to venture into the unknown, to discover what lies beyond the horizon, has shaped the course of history and continues to influence our present and future. The evolution of exploration is a story of curiosity, resilience, and relentless pursuit of knowledge, marked by significant milestones and profound impacts on societies across the globe.

Ancient civilizations were some of the earliest explorers, driven by the need to understand their world and expand their territories. The Egyptians, for example, undertook expeditions along the Nile and into neighboring regions, seeking resources and establishing trade routes. The Phoenicians, revered as master sailors, navigated the Mediterranean Sea, establishing colonies and trade networks that spanned across the ancient world. Their voyages laid the groundwork for future explorations, demonstrating the critical role of navigation in expanding human frontiers.

The Greeks also made significant contributions to early exploration. Figures like Pytheas of Massalia ventured beyond the Pillars of Hercules, exploring the British Isles and possibly even reaching Iceland. These early voyagers were not just adventurers; they were also scholars who documented their findings, contributing to the collective knowledge of their time. Their explorations were driven by a combination of curiosity, commerce, and the desire to map the known world.

The Middle Ages saw a continuation of this exploratory spirit, albeit under different circumstances. The Vikings, with their formidable longships, ventured across the North Atlantic,

reaching as far as North America long before
Columbus. Their expeditions were driven by the
search for new lands to settle and resources to exploit.
The Norse sagas, rich with tales of exploration and
adventure, provide a glimpse into the lives of these
intrepid explorers.

The Crusades, while primarily religious and military
campaigns, also played a role in the evolution of
exploration. European crusaders traveled to the
Middle East, encountering new cultures and trade
goods. These interactions sparked a renewed interest
in the wider world, setting the stage for the Age of
Discovery. The knowledge and technologies acquired
during these encounters, such as improved
shipbuilding techniques and navigational tools, would
prove invaluable in later explorations.

The 15th and 16th centuries marked a pivotal era in
the history of exploration, often referred to as the Age
of Discovery. European nations, driven by the desire
to find new trade routes and expand their empires,
embarked on ambitious voyages across the globe.
Portuguese explorers, under the patronage of Prince
Henry the Navigator, pioneered the exploration of the
West African coast, eventually rounding the Cape of
Good Hope and establishing a sea route to India.
Vasco da Gama's successful voyage to India in 1498
was a monumental achievement, opening up new
avenues for trade and cultural exchange.

Christopher Columbus's transatlantic voyages,
sponsored by Spain, led to the discovery of the New
World in 1492. This event marked the beginning of a
new chapter in global exploration, as European
powers scrambled to explore, conquer, and colonize

the Americas. These expeditions had profound and far-reaching consequences, reshaping the political, economic, and cultural landscapes of the world.

The Age of Discovery was characterized by a blend of adventure and scientific inquiry. Explorers like Ferdinand Magellan, who led the first circumnavigation of the globe, and Sir Francis Drake, who completed his own circumnavigation, expanded the boundaries of the known world. These voyages were fraught with danger and uncertainty, but they also brought back invaluable knowledge and wealth, fueling further exploration and expansion.

The scientific revolution of the 17th and 18th centuries added a new dimension to exploration. Enlightenment thinkers emphasized the importance of empirical observation and scientific inquiry, leading to a more systematic approach to exploration. Figures like Captain James Cook exemplified this new era of scientific exploration. Cook's voyages in the Pacific Ocean, meticulously documented and mapped, contributed significantly to the fields of geography, astronomy, and natural history. His encounters with indigenous peoples also highlighted the complexities and ethical considerations of exploration.

The 19th century saw the rise of exploratory expeditions driven by a combination of scientific curiosity and imperial ambition. European powers, fueled by the Industrial Revolution, embarked on ambitious journeys to chart unknown territories and expand their colonial empires. Explorers like David Livingstone and Henry Morton Stanley ventured into the heart of Africa, mapping uncharted regions and establishing contact with local populations. These expeditions often had profound impacts on the societies they encountered, leading to both cultural exchanges and conflicts.

The exploration of the polar regions during the 19th and early 20th centuries represented some of the most challenging and perilous endeavors in the history of exploration. Explorers like Roald Amundsen, who became the first to reach the South Pole, and Robert Falcon Scott, whose tragic expedition to the same destination captured the imagination of the world, demonstrated the extreme resilience and determination required to explore these harsh environments. These polar expeditions not only expanded our geographical knowledge but also highlighted the limits of human endurance.

The 20th century brought about new frontiers of exploration with the advent of aviation and space travel. Pioneers like Charles Lindbergh, who completed the first solo nonstop transatlantic flight, and Amelia Earhart, who set numerous aviation records, pushed the boundaries of what was possible. Their achievements inspired a new generation of explorers and paved the way for the exploration of the final frontier: space.

The space race of the mid-20th century, driven by geopolitical rivalry between the United States and the Soviet Union, led to some of the most remarkable achievements in the history of exploration. The Soviet Union's launch of Sputnik, the first artificial satellite, in 1957, and Yuri Gagarin's historic flight as the first human in space in 1961, were monumental milestones. The Apollo program, culminating in the Apollo 11 mission that landed humans on the Moon in

1969, represented the pinnacle of human achievement in exploration. These missions expanded our understanding of the cosmos and demonstrated the potential for human ingenuity and perseverance.

In the modern era, exploration continues to evolve, driven by advancements in technology and a renewed sense of curiosity. The exploration of the deep oceans, facilitated by sophisticated submersibles and remotely operated vehicles, has revealed the mysteries of the abyssal plains and the vibrant ecosystems of hydrothermal vents. The discovery of new species and the study of extreme environments have profound implications for our understanding of life on Earth and the potential for life beyond our planet.

As we look to the future, the exploration of Mars and other celestial bodies represents the next frontier. Missions like the Mars rovers and the planned human missions to Mars aim to uncover the secrets of our neighboring planet and assess its potential for future colonization. The search for extraterrestrial life, driven by advances in astrobiology and planetary science, continues to captivate the imaginations of scientists and the public alike.

The evolution of exploration is a testament to the unyielding human spirit and our insatiable desire to expand our horizons. Each era of exploration has built upon the achievements of the past, pushing the boundaries of what is known and possible. The legacy of exploration is reflected in the maps we draw, the knowledge we acquire, and the stories we tell. It is a legacy that continues to inspire future generations to embark on their own journeys of discovery, driven by the same curiosity and determination that have propelled humanity forward for millennia.

Purpose and Structure of the Book

Understanding the purpose and structure of a book is critical for both the writer and the reader. This chapter delves into the essence of why this book exists, its intended impact, and how it is organized to achieve its goals. This approach ensures that readers can navigate the content effectively and extract maximum value from their reading experience.

The primary purpose of this book is to serve as a comprehensive guide for beginners in a particular field. Whether you're stepping into a new career, picking up a hobby, or delving into a complex subject for the first time, this book aims to provide clear, actionable, and insightful information that can help you get started and progress confidently. The idea is to demystify the complexities that often deter beginners and present the material in an accessible and engaging manner.

To achieve this, the book is structured methodically, each chapter building upon the previous one. This logical progression aids in developing a deep understanding of the subject matter. The chapters are not standalone; they are interconnected, creating a cohesive narrative that guides the reader from the basics to more advanced concepts. This structure ensures that as you move forward, the foundation laid in the earlier chapters supports your understanding of the subsequent material.

The book opens with a foundational overview, addressing the fundamental concepts and ideas that are essential for beginners. This sets the stage by providing the necessary background information and context. By establishing a solid foundation, readers can better grasp the more intricate details that follow. This initial groundwork is crucial because it ensures that everyone starts on the same page, regardless of their prior knowledge or experience.

Following this, the chapters dive into specific areas of the subject, each one focusing on a particular aspect. This segmented approach allows for a detailed exploration of each topic, making it easier for readers to absorb the information. Each chapter is designed to be comprehensive yet concise, striking a balance between detail and readability. By breaking down the subject into manageable sections, the book avoids overwhelming the reader and allows for steady, incremental learning.

Throughout the book, real-world examples and case studies are employed to illustrate key points. These examples serve multiple purposes. They not only make the theoretical concepts more relatable and understandable but also demonstrate how these concepts are applied in practical scenarios. This contextual learning is crucial because it bridges the gap between theory and practice, showing readers how to implement what they have learned in real-life situations.

To further enhance the learning experience, the book includes practical exercises and activities. These exercises are designed to reinforce the material covered in each chapter, allowing readers to test their understanding and apply their knowledge. By engaging with the content actively, readers can solidify their learning and gain confidence in their abilities. These activities are varied, catering to different learning styles and preferences, ensuring that everyone can benefit from them.

Additionally, the book emphasizes the importance of reflection and self-assessment. At the end of each chapter, readers are encouraged to reflect on what they have learned and assess their progress. This reflective practice helps consolidate learning, identify areas that need further attention, and set goals for future learning. By making reflection an integral part of the learning process, the book fosters a deeper and more meaningful engagement with the material.

One of the key features of the book is its emphasis on building a solid theoretical foundation while also providing practical, hands-on experience. This dual approach ensures that readers not only understand the underlying principles but also know how to apply them effectively. The theoretical sections are carefully explained, avoiding jargon and overly technical language, while the practical sections provide clear, step-by-step instructions that readers can follow easily.

The book also incorporates insights and advice from experts in the field. These expert contributions add depth and richness to the content, offering readers a diverse range of perspectives and experiences. By learning from those who have already mastered the subject, readers can gain valuable insights and avoid common pitfalls. These expert voices also serve to inspire and motivate, showing readers what is possible with dedication and effort.

To support continuous learning, the book includes resources for further reading and exploration. Each chapter ends with a list of recommended books, articles, websites, and other resources that readers can use to delve deeper into the topics covered. These resources are carefully curated to ensure they are relevant, reliable, and useful. By providing these additional learning materials, the book encourages readers to continue their learning journey beyond its pages.

The conclusion of the book ties everything together, summarizing the key points and reinforcing the main lessons. It also provides guidance on how to continue learning and growing in the subject area. This final chapter serves as both a recap and a springboard, encouraging readers to take what they have learned and apply it in their own lives. It emphasizes the idea that learning is a continuous process and that this book is just the beginning of their journey.

In terms of visual aids, the book includes diagrams, charts, and illustrations where necessary to enhance understanding. These visuals are designed to complement the text, providing a different way of engaging with the material. They are particularly useful for explaining complex concepts or processes, breaking them down into more digestible parts. By combining text and visuals, the book caters to different learning preferences and helps ensure that the material is accessible to all readers.

The writing style of the book is clear, engaging, and conversational. The aim is to make the content as accessible as possible without compromising on accuracy or depth. The tone is friendly and encouraging, creating a supportive learning environment. This approach helps to build a rapport with the reader, making them feel comfortable and motivated to learn.

Acknowledgments

Writing a book is a monumental task, and it is one that cannot be accomplished in isolation. The journey of creating this book has been filled with numerous challenges, triumphs, and invaluable lessons, all of which have been made possible through the support, encouragement, and contributions of many individuals. This chapter is dedicated to acknowledging those who have played a significant role in the creation of this book, offering their time, expertise, and unwavering support.

First and foremost, my deepest gratitude goes to my family. Their constant encouragement and understanding have been the bedrock of this endeavor. To my spouse, whose patience and support have been unwavering, thank you for being my rock. Your belief in my vision and your willingness to shoulder extra responsibilities allowed me the time and space to write. To my children, thank you for your boundless enthusiasm and for understanding when I needed quiet moments to focus.

I would also like to extend my heartfelt thanks to my friends, who have been a source of inspiration and motivation. Your interest in my work and your insightful conversations have often sparked new ideas and perspectives. A special mention goes to those friends who took the time to read early drafts, offering feedback and constructive criticism that helped shape the final manuscript. Your honesty and encouragement have been invaluable.

This book would not have been possible without the guidance and wisdom of my mentors. To my academic advisor, whose profound knowledge and experience in the field have been a guiding light, thank you for your mentorship and for pushing me to strive for excellence. Your feedback on the manuscript was instrumental in refining the ideas and ensuring the clarity and coherence of the content.

A special acknowledgment is owed to the experts and professionals who contributed their insights and experiences to this book. Your willingness to share your knowledge and expertise has enriched the content and provided readers with practical, real-world perspectives. The interviews, case studies, and examples you provided have added depth and authenticity to the material, making it more engaging and relevant.

I am also deeply grateful to my editor, whose meticulous attention to detail and keen eye for quality have significantly improved the manuscript. Your suggestions and edits have enhanced the readability and flow of the book, ensuring that the content is both accessible and informative. Thank you for your patience during the revision process and for your unwavering commitment to excellence.

The support of my publisher has been instrumental in bringing this book to life. To my publishing team, thank you for believing in this project and for your dedication to ensuring its success. Your expertise in production, marketing, and distribution has been crucial in transforming the manuscript into a finished product that can reach a wide audience. I am particularly grateful for the design team, whose creativity and skill have resulted in a visually appealing and professional book.

I would also like to acknowledge the contributions of the researchers and scholars whose work has informed and enriched the content of this book. The vast body of research and literature in the field has provided a solid foundation upon which this book is built. Your dedication to advancing knowledge and understanding has made it possible to present accurate and up-to-date information to readers.

To the numerous colleagues and peers who have provided support and encouragement along the way, thank you for your camaraderie and for the stimulating discussions that have often sparked new ideas. Your feedback and suggestions during conferences, workshops, and informal gatherings have been invaluable in shaping the direction of this book.

I am grateful to the readers and participants who shared their stories and experiences, adding a personal and relatable dimension to the book. Your willingness to open up and share your journeys has provided a rich tapestry of real-life examples that bring the concepts to life. Your stories are a testament to the resilience and determination that are at the heart of the subject matter.

I would be remiss if I did not acknowledge the broader community and organizations that have supported this project. To the libraries, research institutions, and academic bodies that provided access to resources and data, thank you for your invaluable support. The wealth of information available through your collections and databases has been instrumental in conducting thorough research and ensuring the accuracy of the content.

Lastly, I want to express my gratitude to the readers of this book. Your interest and engagement are what make this endeavor worthwhile. It is my hope that this book provides you with the knowledge, insights, and inspiration you seek. Thank you for embarking on this journey with me, and I hope that you find the content as enriching and valuable as it was for me to create.

Chapter 2
The Age of Discovery

The Dawn of Global Exploration

The dawn of global exploration marks one of the most transformative periods in human history. It was an era where curiosity, ambition, and the desire for new trade routes and knowledge propelled nations into uncharted waters. This chapter delves into the motivations, key figures, and significant events that defined this remarkable period, highlighting how these explorations reshaped the world.

Driven by a desire for spices, silk, and other luxurious goods, European nations sought new trade routes to Asia. The overland routes were long, arduous, and controlled by various intermediaries, making goods excessively expensive. Navigators and monarchs envisioned sea routes as a solution, spurring a wave of explorations.

Portugal emerged as a pioneer, thanks to the efforts of Prince Henry the Navigator. Though he never sailed himself, his patronage of explorers and navigators laid the groundwork for future voyages. By establishing a school for navigation and funding numerous expeditions along the African coast, Henry set the stage for Portugal's maritime dominance.

In 1488, Bartolomeu Dias rounded the Cape of Good Hope, proving that the Atlantic and Indian Oceans were connected. This breakthrough opened the sea route to Asia, which Vasco da Gama successfully navigated in 1498, reaching India and establishing a vital trade link. Portugal's success inspired other European powers to pursue their own exploratory ventures.

Spain, motivated by the same economic and competitive pressures, backed Christopher Columbus's ambitious plan to reach Asia by sailing west. In 1492, Columbus's expedition landed in the Caribbean, mistakenly identifying it as the East Indies. This discovery, though not the direct route to Asia he had hoped for, was monumental. It heralded the beginning of European colonization in the Americas and initiated a wave of exploration and conquest.

The Treaty of Tordesillas in 1494, brokered by the Pope, divided the newly discovered lands outside Europe between Portugal and Spain along a meridian 370 leagues west of the Cape Verde islands. This agreement aimed to resolve conflicts over newly explored territories and solidified the two nations' spheres of influence.

Other European nations, such as England, France, and the Netherlands, soon joined the race for exploration and colonization. John Cabot, an Italian navigator under the commission of England, reached the coast of North America in 1497, laying the groundwork for England's later claims in the New World. France's Jacques Cartier explored the St. Lawrence River in the 1530s, establishing a French presence in Canada. The Dutch, renowned for their maritime prowess, established trade routes to the East Indies, challenging Portuguese dominance.

The age of exploration was not solely about new trade routes and territorial expansion. It was also a period of significant scientific and cultural exchanges. Explorers brought back not only goods but also knowledge about geography, biology, and astronomy. These discoveries expanded the European worldview and led to the development of cartography and navigation techniques.

However, this era also had profound and often devastating impacts on the indigenous populations of the discovered lands. The arrival of Europeans brought diseases to which the native peoples had no immunity, leading to catastrophic population declines. The conquest and colonization efforts often resulted in the exploitation and oppression of indigenous peoples. The transatlantic slave trade emerged during this period, with Africans forcibly taken to the New World to work in plantations and mines.

One of the most significant figures of this period was Ferdinand Magellan, whose expedition (1519-1522) achieved the first circumnavigation of the Earth. Although Magellan himself did not survive the journey, his fleet's successful return to Spain proved definitively that the world was round and that it was possible to sail around it. This monumental achievement underscored the vastness of the globe and the interconnectedness of its oceans.

The era of global exploration also saw the establishment of the first global trading networks. The Portuguese and Spanish empires created extensive trade routes connecting Europe, Africa, Asia, and the Americas. The Manila Galleons, Spanish trading ships, sailed across the Pacific, linking the Philippines with Mexico and facilitating the exchange of goods and cultures across the Pacific Ocean.

As explorers charted new territories, they also encountered diverse cultures and civilizations. The exchange of ideas, technologies, and goods between Europe and the rest of the world fundamentally transformed societies on both sides. European diets were enriched by the introduction of new crops such as potatoes, tomatoes, and maize, while European goods and technologies found their way into indigenous cultures.

The legacy of the age of exploration is multifaceted. It laid the foundations for the modern globalized world, with its intricate network of trade routes and cultural exchanges. It also set the stage for the colonial empires that would dominate global politics and economics for centuries. The knowledge gained during this period fueled the scientific revolution and the Enlightenment, driving further advances in navigation, astronomy, and geography.

However, it is essential to recognize the darker aspects of this legacy. The exploitation and destruction of indigenous cultures, the establishment of the transatlantic slave trade, and the environmental impacts of colonization are all part of this complex history. Understanding the full scope of the age of exploration requires acknowledging both its achievements and its consequences.

In reflecting on this era, one must admire the courage and determination of the explorers who ventured into the unknown, often at great personal risk. Their voyages required not only advanced navigational skills and seafaring technology but also immense bravery and resilience. These explorers were driven by a mix of ambition, curiosity, and the desire for wealth and glory, motivations that resonate throughout human history.

The dawn of global exploration stands as a testament to the human spirit's insatiable curiosity and drive for discovery. It reshaped the world in ways that are still felt today, fostering connections between distant lands and peoples. As we look back on this period, we can draw lessons about the power of exploration and the importance of approaching new frontiers with respect and responsibility.

Notable Explorers of the Era

The era of exploration was defined by the daring adventurers who risked their lives to discover new lands, establish trade routes, and expand the horizons of human knowledge. These explorers, driven by a mix of curiosity, ambition, and the desire for glory, left indelible marks on history. Their journeys not only mapped the world but also reshaped the global landscape, setting the stage for the interconnected world we live in today.

Among the most notable explorers of this era was Christopher Columbus, an Italian navigator whose 1492 voyage, under the auspices of the Spanish Crown, led to the European discovery of the New World. Columbus's journey across the Atlantic was fraught with uncertainty and peril. Believing he could reach Asia by sailing west, Columbus instead landed in the Caribbean, opening the door to the colonization and exploration of the Americas. His voyages, although based on a geographical miscalculation, were pivotal in connecting Europe with the Americas.

Another significant figure was Vasco da Gama, a Portuguese explorer who successfully navigated a sea route from Europe to India in 1498. This monumental journey around the Cape of Good Hope not only established a direct maritime link between Europe and Asia but also marked the beginning of a new era of global trade. Da Gama's voyages were instrumental in establishing Portugal as a dominant maritime power and in securing lucrative spice trade routes that bypassed the overland routes controlled by Ottoman and Arab traders.

Ferdinand Magellan, a Portuguese explorer sailing under the Spanish flag, is best known for his expedition that achieved the first circumnavigation of the Earth. Although Magellan himself did not complete the entire journey—he was killed in the Philippines—his fleet, led by Juan Sebastián Elcano after his death, returned to Spain in 1522. This expedition conclusively demonstrated the vastness of the Earth and the interconnectedness of its oceans, forever changing the European understanding of the world.

John Cabot, an Italian navigator commissioned by England, played a crucial role in the exploration of North America. His 1497 voyage to the coast of what is now Canada laid the groundwork for England's later claims in the New World. Cabot's journeys were among the earliest European explorations of the North American continent, and his reports of rich fishing grounds off the coast of Newfoundland spurred further interest and exploration by European powers.

Jacques Cartier, a French explorer, made significant contributions to the exploration of Canada. In the 1530s, Cartier explored the Gulf of St. Lawrence and the St. Lawrence River, laying the foundation for French claims in North America. His interactions with indigenous peoples and his detailed accounts of the region provided valuable information that would guide future explorers and settlers.

Henry Hudson, an English explorer, undertook several voyages in search of a northwest passage to Asia. Although he did not find the elusive passage, Hudson's explorations of the Arctic and northeastern North America were significant. His voyages led to the mapping of Hudson Bay and Hudson River, areas that would later become central to the fur trade and European colonization efforts.

The Dutch explorer Abel Tasman is renowned for his discoveries in the southern hemisphere. Commissioned by the Dutch East India Company, Tasman's voyages in the 1640s led to the European discovery of Tasmania, New Zealand, and the Fiji Islands. His explorations expanded European knowledge of the Pacific region and contributed to the Dutch dominance in the spice trade.

James Cook, a British explorer of the 18th century, conducted three major voyages that greatly expanded European knowledge of the Pacific Ocean and its islands. Cook's expeditions mapped large parts of the Pacific, including the east coast of Australia, New

Zealand, and the Hawaiian Islands. His meticulous mapping and scientific observations made significant contributions to the fields of geography, astronomy, and natural history.

The Spanish explorer Hernán Cortés is infamous for his conquest of the Aztec Empire in present-day Mexico. Arriving in 1519, Cortés's military campaigns, alliances with indigenous groups, and strategic use of advanced weaponry led to the fall of Tenochtitlán in 1521. Cortés's actions had profound and lasting impacts on the indigenous populations and the course of history in the Americas.

Similarly, Francisco Pizarro, another Spanish conquistador, led the expedition that conquered the Inca Empire in South America. Landing in 1532, Pizarro's small force captured the Inca ruler Atahualpa and eventually took control of the vast Inca territory. The conquest of Peru brought immense wealth to Spain and played a key role in the expansion of the Spanish Empire in the New World.

Other explorers, such as Samuel de Champlain, known as the "Father of New France," were instrumental in the exploration and colonization of Canada. Champlain's voyages in the early 1600s led to the establishment of Quebec City and the mapping of the Great Lakes region. His efforts to foster alliances with indigenous peoples and his detailed records of the geography and cultures of the region were crucial to the success of French colonial endeavors.

The era of exploration was also marked by the contributions of lesser-known but equally important figures. Bartolomeu Dias, for instance, was the first

European to sail around the southern tip of Africa, the Cape of Good Hope, in 1488. This voyage proved that a sea route to Asia was possible and paved the way for future explorers like Vasco da Gama.

Pedro Álvares Cabral, another Portuguese navigator, is credited with the discovery of Brazil in 1500. His fleet, intended for India, was blown off course, leading to the unexpected landfall on the coast of South America. Cabral's discovery secured Portugal's claim to Brazil, which would become a major part of its colonial empire.

These explorers, driven by a mix of ambition, curiosity, and the pursuit of wealth, charted new territories and connected distant parts of the globe. Their voyages were fraught with dangers, including uncharted waters, hostile encounters with indigenous peoples, and the constant threat of disease. Yet, their determination and resilience expanded the boundaries of the known world and laid the foundations for the modern age of global interaction.

The impact of these explorers extended beyond mere geographical discoveries. They facilitated the exchange of goods, ideas, and cultures across continents, leading to significant economic, social, and political changes. The introduction of new crops, animals, and technologies transformed societies on both sides of the Atlantic and beyond. However, these exchanges also had devastating consequences for indigenous populations, including displacement, disease, and cultural disruption.

Technological Advancements and Their Impact

Technological advancements have been the driving force behind human progress for centuries, shaping societies and economies in profound ways. From the invention of the wheel to the digital revolution, each significant leap in technology has redefined the boundaries of what is possible, transforming industries, lifestyles, and global interactions.

The Industrial Revolution, which began in the late 18th century, marked a pivotal moment in history. It was characterized by the transition from agrarian economies to industrialized ones, driven by innovations such as the steam engine, mechanized textile production, and the development of iron and steel manufacturing. The steam engine, invented by James Watt, revolutionized transportation and manufacturing, enabling factories to operate more efficiently and facilitating the movement of goods and people over long distances. This period also saw the rise of the factory system, which centralized production and led to the growth of urban centers.

The impact of these advancements was profound. Economies grew rapidly, and new job opportunities emerged in urban areas, drawing people from rural regions. However, this shift also brought about significant social changes and challenges. Working conditions in factories were often harsh, with long hours and dangerous environments, leading to the rise of labor movements and calls for better working conditions and rights. The Industrial Revolution also had environmental consequences, as the increased use of coal and other fossil fuels led to pollution and the depletion of natural resources.

The 19th century witnessed further technological progress with the advent of electricity and the internal combustion engine. Thomas Edison and Nikola Tesla were instrumental in developing electrical power systems, which transformed everyday life by providing a reliable source of energy for homes, businesses, and factories. The internal combustion engine, perfected by inventors like Karl Benz and Henry Ford, revolutionized transportation by making automobiles more accessible to the general public. Ford's assembly line production method significantly lowered the cost of manufacturing cars, making them affordable for many people and changing the landscape of cities and towns.

Communication technologies also saw remarkable advancements during this period. The telegraph, developed by Samuel Morse, and later the telephone, invented by Alexander Graham Bell, revolutionized the way people communicated over long distances. These innovations enabled faster and more efficient communication, which was crucial for business operations, military strategy, and personal interactions. The establishment of global telegraph networks and undersea cables connected continents and facilitated international trade and diplomacy.

The 20th century brought about a wave of technological innovations that further transformed society. The development of radio and television changed the way information was disseminated and consumed, creating a new era of mass communication. Radio became a powerful tool for news, entertainment, and propaganda, while television brought visual media into people's homes, shaping public opinion and culture.

The mid-20th century saw the dawn of the computer age, with pioneers like Alan Turing and John von Neumann laying the groundwork for modern computing. The invention of the transistor by John Bardeen, Walter Brattain, and William Shockley in 1947 revolutionized electronics by allowing for the creation of smaller, more efficient devices. This led to the development of the first computers, which, although large and expensive, began to revolutionize data processing and business operations.

As computers became more advanced and affordable, they started to permeate various aspects of life. The introduction of personal computers in the 1970s and 1980s, spearheaded by companies like Apple and IBM, brought computing power to individuals and small businesses. This democratization of technology spurred innovation and creativity, leading to the development of software that could perform complex tasks and solve problems in new ways.

The late 20th century and early 21st century have been defined by the rise of the internet and digital technologies. The internet, initially developed as a military communication network, quickly evolved into a global information superhighway. The creation of the World Wide Web by Tim Berners-Lee in 1989 made it possible for people to access and share information easily, leading to the explosion of online content and services. E-commerce, social media, and digital communication platforms have transformed how people shop, interact, and conduct business.

Mobile technology has further accelerated these changes. The introduction of smartphones, with their powerful processors and constant internet connectivity, has put the world at people's fingertips. Mobile apps have revolutionized industries such as banking, healthcare, entertainment, and transportation. The gig economy, facilitated by platforms like Uber and Airbnb, has created new opportunities and challenges, reshaping traditional employment and business models.

Technological advancements have also had a profound impact on healthcare. Medical devices and diagnostic tools have become more sophisticated, enabling earlier and more accurate diagnoses. Advances in biotechnology and pharmaceuticals have led to new treatments and cures for diseases that were once considered untreatable. Telemedicine has made healthcare more accessible, allowing patients to consult with doctors remotely and receive care from the comfort of their homes.

In agriculture, technology has revolutionized the way food is produced and distributed. Mechanization, improved irrigation techniques, and the development of high-yield crop varieties have increased food production and efficiency. Precision agriculture, which uses GPS technology, sensors, and data analytics, allows farmers to optimize their use of resources and maximize crop yields while minimizing environmental impact.

While the benefits of technological advancements are undeniable, they also pose significant ethical and societal challenges. The rapid pace of change can lead to job displacement, as automation and artificial intelligence take over tasks previously performed by humans. This necessitates a focus on education and retraining to ensure that workers can adapt to new roles and industries.

Privacy and security are also major concerns in the digital age. The vast amounts of data generated by digital technologies can be vulnerable to breaches and misuse, raising questions about how to protect individuals' personal information and ensure cybersecurity. The rise of surveillance technologies and the potential for misuse by governments and corporations further complicate these issues.

Environmental sustainability is another critical consideration. While technology has the potential to address environmental challenges, such as through the development of renewable energy sources and more efficient resource management, it can also contribute to environmental degradation. The production and disposal of electronic devices, for instance, create significant amounts of waste and pollution.

The role of technology in society is a complex and evolving issue that requires careful consideration and balanced approaches. Policymakers, businesses, and individuals must work together to harness the benefits of technological advancements while mitigating their negative impacts. This involves investing in education and training, developing robust regulatory frameworks, and fostering a culture of ethical innovation.

The Race for New Lands

The race for new lands has been a defining aspect of human history, driven by a desire for exploration, conquest, and expansion. This pursuit, which began with early migrations and continued through the age of exploration, has shaped the political, social, and economic landscapes of nations around the globe. The quest for new territories often involved complex interactions between explorers, indigenous populations, and competing powers, resulting in a tapestry of stories that reveal both the heights of human ingenuity and the depths of human ambition.

The earliest migrations of human populations can be traced back to our ancestors' movement out of Africa approximately 60,000 years ago. These migrations were driven by a combination of factors, including climate changes and the search for food and resources. As groups of early humans ventured into new territories, they encountered diverse environments and adapted to a variety of ecological niches. These migrations laid the groundwork for the peopling of continents and the development of distinct cultures and societies.

One of the most significant periods in the race for new lands was the Age of Exploration, which spanned the late 15th to the early 17th centuries. This era was marked by European powers seeking new trade routes, territories, and resources. The voyages of explorers such as Christopher Columbus, Vasco da Gama, and Ferdinand Magellan opened up new frontiers and established the basis for European colonial empires. Columbus's 1492 voyage, sponsored by Spain, led to the European discovery of the

Americas, an event that had monumental consequences for both the Old and New Worlds.

The race for new lands during the Age of Exploration was fueled by a combination of economic, political, and religious motivations. European nations sought to expand their wealth and influence by establishing colonies and tapping into the resources of newly discovered territories. The desire for precious metals, spices, and other valuable commodities drove explorers to undertake perilous journeys across uncharted oceans. Additionally, the spread of Christianity was a significant motivation, with missionaries accompanying explorers to convert indigenous populations.

The impact of European exploration and colonization on indigenous peoples was profound and often devastating. The arrival of Europeans brought diseases to which native populations had no immunity, leading to catastrophic population declines. In many cases, indigenous societies were subjected to violence, displacement, and forced labor. The Columbian Exchange, the widespread transfer of plants, animals, and diseases between the Old and New Worlds, reshaped ecosystems and economies on both sides of the Atlantic.

As European powers established colonies around the world, competition for new lands intensified. The 16th and 17th centuries saw numerous conflicts over territory, often involving complex alliances and rivalries among European nations. The Treaty of Tordesillas in 1494, brokered by the Pope, attempted to divide the newly discovered lands between Spain and Portugal. However, other nations such as

England, France, and the Netherlands soon entered the fray, challenging Spanish and Portuguese dominance.

The race for new lands was not limited to the Americas. European powers also sought to expand their influence in Africa, Asia, and the Pacific. The Portuguese established a trading empire along the coast of Africa and in the Indian Ocean, while the Dutch East India Company dominated trade in Southeast Asia. The British and French established colonies in India, leading to a protracted struggle for control of the subcontinent. In the Pacific, explorers such as James Cook charted new territories and claimed lands for their respective nations.

The race for new lands continued into the 19th and early 20th centuries, culminating in the era of New Imperialism. During this period, European powers, along with the United States and Japan, engaged in a frenzied scramble for colonies and spheres of influence. The Berlin Conference of 1884-1885 formalized the partition of Africa, with European nations carving up the continent with little regard for existing ethnic and cultural boundaries. This period also saw the colonization of Southeast Asia, the Pacific Islands, and parts of China.

The motivations behind New Imperialism were multifaceted. Industrialization had created a demand for raw materials and new markets, driving nations to seek control over resource-rich territories. Nationalism and the desire for prestige also played a role, as colonial possessions were seen as a measure of a nation's power and influence. Additionally, the spread of Western ideologies, including the belief in the civilizing mission, justified the expansion of empires and the subjugation of non-Western peoples.

The impact of New Imperialism on colonized regions was profound and often traumatic. Colonial rule frequently involved the exploitation of local populations and resources, leading to economic and social disruption. Traditional societies were often dismantled, and indigenous cultures were suppressed or marginalized. However, colonial rule also brought about significant infrastructural developments, such as the construction of railways, roads, and ports, which facilitated the integration of colonies into the global economy.

Resistance to colonial rule emerged in various forms, from armed rebellions to intellectual and political movements. Indigenous leaders and activists sought to reclaim their lands and assert their rights, often drawing on traditional cultural practices and values. The struggle for independence gained momentum in the 20th century, leading to the decolonization of Africa, Asia, and the Caribbean. The end of colonial rule marked a significant shift in global geopolitics, as newly independent nations sought to assert their sovereignty and navigate the complexities of the Cold War era.

The race for new lands did not end with the decolonization process. In the latter half of the 20th century, technological advancements opened up new frontiers, including space exploration and the deep sea. The Space Race between the United States and the Soviet Union, exemplified by the Apollo moon landings, represented a new kind of territorial competition, driven by scientific curiosity and geopolitical rivalry. The exploration of the deep sea, facilitated by advances in submersible technology, has revealed new ecosystems and resources, raising questions about the sustainable use of these largely uncharted territories.

In the contemporary era, the race for new lands has taken on new dimensions. Climate change and environmental degradation are reshaping the geopolitical landscape, as nations compete for access to dwindling resources and habitable land. The melting of polar ice caps has opened up new shipping routes and access to untapped natural resources in the Arctic, leading to increased interest and competition among Arctic nations. Additionally, the search for sustainable energy sources has spurred interest in renewable energy projects, such as offshore wind farms and solar arrays in desert regions.

Cultural Exchanges and Conflicts

Human history is a tapestry woven from countless threads of cultural exchanges and conflicts. These interactions have shaped civilizations, influencing languages, beliefs, technologies, and social structures. The encounters between different cultures have often led to periods of remarkable growth and understanding, but they have also sparked conflicts that have left lasting scars. Understanding the dynamics of cultural exchanges and conflicts is essential for appreciating the complex nature of our global society.

One of the earliest and most significant instances of cultural exchange occurred along the Silk Road. This ancient network of trade routes connected the East and West, facilitating the exchange of goods, ideas, and technologies. The Silk Road was not a single road but a series of interconnected trade paths that spanned from China to the Mediterranean. Merchants, pilgrims, and diplomats traveled these routes, bringing silk, spices, precious metals, and other goods to distant markets. Alongside these material exchanges, there was a flow of knowledge and cultural practices. For example, Buddhism spread from India to China, Korea, and Japan, profoundly influencing the spiritual and cultural landscapes of these regions.

The Silk Road also exemplified the complexities of cultural interaction. While it fostered trade and communication, it was also a conduit for conflict and disease. Different empires and tribes sought to control sections of the route, leading to skirmishes and battles. Additionally, the movement of people and goods facilitated the spread of diseases such as the bubonic plague, which had devastating effects on populations across Eurasia.

The Age of Exploration marked another critical period of cultural exchanges and conflicts. As European explorers set sail to discover new lands, they encountered diverse cultures in Africa, Asia, and the Americas. These encounters were often marked by a mix of curiosity and conquest. The arrival of Europeans in the Americas, for instance, led to the Columbian Exchange, a massive transfer of plants, animals, and diseases between the Old and New Worlds. European settlers introduced wheat, horses, and cattle to the Americas, while crops like maize, potatoes, and tomatoes were brought to Europe. This exchange transformed diets and agricultural practices on both sides of the Atlantic.

However, these encounters were not without conflict. The European colonization of the Americas led to the displacement, and often the destruction, of indigenous cultures. The introduction of diseases like smallpox decimated native populations, who had no immunity to these foreign illnesses. Conflicts over land and resources were frequent, as European powers sought to establish dominance in the New World. The brutality of conquest, enslavement, and forced conversion left deep scars on the cultural fabric of indigenous societies.

The transatlantic slave trade is another somber example of cultural exchange and conflict. Over the course of more than three centuries, millions of Africans were forcibly taken from their homelands and transported to the Americas. This horrific trade was driven by European demand for labor to work on plantations in the New World. The cultural impact of the slave trade was profound and multifaceted. African cultures were disrupted, families torn apart, and entire societies traumatized. Yet, the resilience of enslaved Africans led to the creation of rich, syncretic cultures in the Americas. African traditions in music, dance, religion, and cuisine blended with European and indigenous influences, giving rise to vibrant new cultural forms that continue to thrive today.

The colonial era also saw significant cultural exchanges between Europe and Asia. The British colonization of India, for example, led to profound cultural interactions. British administrators, soldiers, and merchants brought Western education, legal systems, and technologies to India, while Indian art, literature, and philosophy began to influence European thought. This period also saw the rise of Indian nationalism, as educated Indians began to challenge colonial rule and assert their cultural identity. Figures like Mahatma Gandhi and Rabindranath Tagore drew on both Indian and Western traditions to advocate for independence and social reform.

In the 20th century, two world wars and the subsequent decolonization process further reshaped cultural exchanges and conflicts. The world wars brought people from diverse backgrounds into contact as soldiers, workers, and refugees moved across borders. These interactions led to a greater awareness of different cultures and sometimes to friction. The aftermath of the wars saw the dismantling of colonial empires and the emergence of new nations. The newly independent countries often faced the challenge of forging a national identity that reconciled colonial legacies with indigenous traditions.

The Cold War era introduced another dimension to cultural exchanges and conflicts. The ideological struggle between the United States and the Soviet Union was not only a political and military contest but also a cultural one. Both superpowers sought to project their way of life as superior, leading to a competition in arts, sports, and education. The exchange programs, propaganda efforts, and cultural diplomacy of this period aimed to win hearts and minds across the globe. This era also saw significant migration, as people fled political oppression or sought better opportunities, leading to multicultural societies particularly in the United States and Western Europe.

In recent decades, globalization has accelerated cultural exchanges on an unprecedented scale. Advances in communication and transportation have made it easier for people to share ideas, traditions, and innovations. The internet, in particular, has created a global village where cultures can interact instantaneously. Social media platforms enable the sharing of music, fashion, and cuisine across borders, fostering a greater appreciation of cultural diversity.

Yet, globalization has also led to new conflicts. The rapid spread of Western consumer culture has been met with resistance in some parts of the world, where people fear the erosion of their traditional values. Economic inequalities and the displacement of local industries by global corporations have fueled anti-globalization sentiments. The clash between global and local cultures is evident in debates over cultural appropriation, immigration, and national identity.

One of the most significant challenges in contemporary cultural exchanges is striking a balance between embracing diversity and maintaining cultural integrity. Cultural exchanges can enrich societies, fostering creativity and innovation. However, they can also lead to homogenization, where unique cultural identities are lost in a sea of global influences. Preserving cultural heritage while engaging with the broader world requires thoughtful policies and practices that respect and celebrate diversity.

Education plays a crucial role in promoting positive cultural exchanges. By teaching young people about different cultures and histories, we can foster mutual understanding and respect. Cultural exchange programs, such as student exchanges and international internships, provide valuable opportunities for people to experience different ways of life firsthand. These experiences can break down stereotypes and build bridges between communities.

Chapter 3

Navigating the Unknown

The Challenges of Early Navigation

Navigating the vast and unpredictable oceans has always been a formidable challenge for humanity. In the early days of exploration, mariners faced a myriad of obstacles that tested their ingenuity, courage, and resilience. The challenges of early navigation were numerous, ranging from the limitations of technology to the perils of the natural world. Understanding these challenges provides insight into the incredible achievements of early explorers and the advancements that revolutionized maritime travel.

One of the primary challenges of early navigation was the lack of accurate maps and charts. Early explorers often relied on rudimentary maps that were based on limited knowledge and hearsay. These maps were not only incomplete but frequently inaccurate, leading to dangerous miscalculations. The absence of detailed and reliable charts made it difficult for sailors to plot their courses with any degree of certainty. This lack of precise information often resulted in ships straying off course, sometimes leading to disastrous consequences.

The instruments available to early navigators were also relatively primitive. The magnetic compass, one of the most crucial tools for navigation, had been known since ancient times, but its use in maritime travel became widespread only in the Middle Ages.

Even then, the compass was not always reliable due to magnetic variations and the interference caused by the iron on ships. The astrolabe and the cross-staff were used to determine latitude by measuring the angle of celestial bodies above the horizon, but these instruments required clear skies and a skilled hand. Determining longitude, however, remained a significant challenge until the invention of the marine chronometer in the 18th century.

Weather was another formidable obstacle for early navigators. Sailors had to contend with unpredictable and often violent weather conditions, including storms, high winds, and rough seas. Without modern meteorological knowledge and forecasting tools, early mariners had little warning of impending weather changes. Storms could drive ships off course, damage or destroy vessels, and endanger the lives of the crew. The lack of reliable weather information meant that sailors had to rely on their experience and intuition, which were not always sufficient to avoid disaster.

The physical condition of the ships themselves posed another significant challenge. Early ships were often made of wood and were susceptible to damage from the elements, as well as from collisions with rocks and other vessels. The construction techniques of the time limited the size and durability of ships, making long voyages particularly perilous. Leaks, structural weaknesses, and the constant threat of shipwreck were ever-present concerns for early navigators.

One of the most daunting challenges faced by early sailors was the threat of disease. Life aboard ship was harsh and unsanitary, with limited access to fresh food and clean water. Scurvy, caused by a deficiency of vitamin C, was a common and deadly affliction among sailors on long voyages. The cramped and unhygienic conditions on ships also facilitated the spread of other diseases, such as typhus and dysentery. The high mortality rate among crews was a major obstacle to successful navigation and exploration.

Navigating unknown and uncharted waters was fraught with dangers. Early explorers had to be constantly vigilant for hidden reefs, sandbanks, and other hazards that could spell disaster for their ships. The lack of detailed knowledge about these waters made it difficult to avoid such dangers, and many ships were lost to these unseen perils. Additionally, encounters with hostile indigenous peoples or rival explorers could lead to violent confrontations, adding another layer of risk to maritime exploration.

Despite these daunting challenges, early navigators made remarkable achievements that paved the way for future exploration and trade. The voyages of explorers like Christopher Columbus, Vasco da Gama, and Ferdinand Magellan are testaments to human perseverance and ingenuity. Columbus's journey across the Atlantic in 1492, for instance, was a monumental achievement given the limited knowledge and technology of the time. His voyage not only opened up the New World to European exploration but also demonstrated the feasibility of transoceanic travel.

The Portuguese explorer Vasco da Gama's successful navigation around the Cape of Good Hope to reach India in 1498 was another landmark achievement. This voyage established a direct sea route between

Europe and Asia, revolutionizing global trade and commerce. Da Gama's success was made possible by advancements in navigational techniques and the accumulation of knowledge from earlier explorers.

Ferdinand Magellan's expedition, which resulted in the first circumnavigation of the globe, stands as one of the most significant achievements in the history of navigation. Despite the loss of Magellan himself during the journey, his fleet's successful completion of the voyage in 1522 provided invaluable information about the world's geography and demonstrated the vast potential of maritime exploration.

The advancements in shipbuilding and navigational technology that followed these early voyages were instrumental in overcoming many of the challenges faced by early navigators. The development of the caravel, a small, highly maneuverable ship with a distinctive lateen sail, allowed explorers to sail closer to the wind and navigate more challenging waters. The refinement of navigational instruments, such as the sextant and the marine chronometer, greatly improved the accuracy of maritime navigation.

The establishment of naval academies and the systematic training of sailors also played a crucial role in advancing navigational knowledge and techniques. These institutions provided rigorous training in the use of navigational instruments, celestial navigation, and seamanship, equipping sailors with the skills needed to undertake long and complex voyages.

The development of more accurate and detailed maps and charts was another critical factor in overcoming the challenges of early navigation. The work of cartographers like Gerardus Mercator, who created the Mercator projection, revolutionized mapmaking by providing a way to represent the curved surface of the Earth on a flat map with minimal distortion. This innovation made it easier for navigators to plot their courses and navigate with greater precision.

The challenges of early navigation also led to significant cultural and scientific exchanges. The need for better navigational knowledge spurred the collection and dissemination of information from diverse sources. Explorers brought back not only geographical knowledge but also information about the peoples, plants, and animals they encountered. This exchange of knowledge contributed to the growth of scientific understanding and the development of new technologies.

The Role of Maps and Cartography

Maps have long been essential tools for understanding and navigating the world. From ancient parchment scrolls to digital screens, cartography has evolved significantly, shaping our perception of geography and enabling exploration and discovery. The role of maps and cartography extends far beyond simple navigation; they are instruments of power, tools of knowledge, and expressions of art. For beginners, understanding the history, development, and impact of cartography provides a foundation for appreciating these intricate representations of our world.

In ancient times, maps were often rudimentary, created with limited knowledge and materials. Early cartographers relied on the accounts of travelers, sailors, and merchants to piece together the known world. One of the oldest known maps, the Babylonian World Map, dates back to the 6th century BCE and depicts a simplified view of the world with Babylon at its center. This map, while not accurate by modern standards, illustrates the early human desire to understand and document the surrounding environment.

The Greeks made significant advancements in cartography, with figures like Anaximander and Ptolemy leading the way. Anaximander, in the 6th century BCE, is credited with creating one of the earliest known world maps that attempted to represent the earth's surface proportionally. Ptolemy's work in the 2nd century CE laid the groundwork for modern cartography with his book, "Geographia," which included a compilation of geographical knowledge and instructions for creating maps. His use of a coordinate system to map the known world was revolutionary and influenced cartographers for centuries.

During the Middle Ages, European cartography experienced a decline, but it thrived in the Islamic world. Islamic scholars translated and preserved Greek and Roman geographical texts, expanding upon them with their own observations and travels. The maps created during this period, such as the Tabula Rogeriana by al-Idrisi in the 12th century, were highly detailed and accurate for their time. These maps not only aided in navigation but also facilitated trade and cultural exchange between different regions.

The Age of Exploration in the 15th and 16th centuries marked a significant turning point in the history of cartography. European explorers, driven by the desire for new trade routes and territorial expansion, embarked on voyages that vastly expanded the known world. Cartographers like Gerardus Mercator and Abraham Ortelius played pivotal roles in this era. Mercator's 1569 world map introduced the Mercator projection, which allowed for more accurate navigation by representing lines of constant course, or rhumb lines, as straight segments. Although this projection distorts size and shape near the poles, it became invaluable for maritime navigation.

Ortelius, on the other hand, is known for creating the first modern atlas, "Theatrum Orbis Terrarum," in 1570. This atlas compiled maps from various sources, providing a comprehensive view of the world. The work of these cartographers not only advanced geographical knowledge but also facilitated exploration, colonization, and the spread of European influence across the globe.

As exploration continued, the demand for more precise and detailed maps grew. The 18th and 19th centuries saw significant advancements in surveying techniques and cartographic tools. The invention of the chronometer by John Harrison in the 18th century allowed for accurate determination of longitude, a critical factor in creating precise maps. Triangulation, a method of determining the location of a point by measuring angles to it from known points, became a standard technique in surveying.

The development of printing technology also revolutionized cartography. Maps could be produced more quickly and distributed widely, making geographical knowledge accessible to a broader audience. The work of cartographers like James Cook, who meticulously charted the coastlines of Newfoundland, Australia, and New Zealand, contributed to the accuracy and detail of maps during this period. Cook's maps were widely regarded for their precision and greatly expanded the known geography of the time.

In the 20th century, technological advancements continued to transform cartography. Aerial photography, satellite imagery, and geographic information systems (GIS) revolutionized the way maps were created and used. Aerial photography provided a bird's-eye view of the landscape, allowing for the creation of highly detailed topographic maps. Satellite imagery further enhanced the accuracy and scope of cartographic data, enabling the mapping of previously inaccessible areas.

GIS technology allowed for the integration and analysis of various types of geographical data, from topography and climate to population and infrastructure. This technology not only improved the accuracy and utility of maps but also opened up new possibilities for spatial analysis and decision-making. GIS has become an essential tool in fields ranging from urban planning and environmental management to disaster response and public health.

The role of maps and cartography in contemporary society extends beyond navigation and exploration. Maps are powerful tools for communication and storytelling, conveying complex information in a visual and accessible format. They are used in education to teach geography and history, in journalism to illustrate news stories, and in art to create visually stunning representations of the world.

Cartography also plays a crucial role in addressing global challenges. Climate change, for example, requires detailed and accurate maps to monitor changes in the environment, such as rising sea levels, shifting weather patterns, and the distribution of natural resources. Maps are essential for disaster response and management, providing critical information for planning and coordination in the aftermath of natural disasters like earthquakes, hurricanes, and wildfires.

In the digital age, maps have become more interactive and dynamic. Online mapping platforms like Google Maps and OpenStreetMap allow users to explore the world from their devices, providing real-time navigation, traffic updates, and location-based services. These platforms rely on vast amounts of data collected from various sources, including satellites, sensors, and user contributions, to create accurate and up-to-date maps.

Crowdsourced mapping projects have also emerged, harnessing the power of collective knowledge and participation. OpenStreetMap, for example, relies on contributions from volunteers worldwide to create and update its maps. This collaborative approach has proven invaluable in rapidly mapping areas affected by disasters or in regions where traditional mapping efforts are limited.

As we look to the future, the role of maps and cartography will continue to evolve. Advances in technology, such as augmented reality (AR) and virtual reality (VR), offer new possibilities for immersive and interactive mapping experiences. AR applications can overlay digital information onto the physical world, enhancing navigation and exploration. VR can create fully immersive environments for virtual exploration and spatial analysis.

Moreover, the integration of artificial intelligence and machine learning in cartography promises to further enhance the accuracy, efficiency, and capabilities of map-making. These technologies can analyze vast amounts of data, identify patterns, and generate highly detailed and dynamic maps that adapt to changing conditions.

Pioneering Navigators

Navigating the uncharted waters of the world has always been a blend of bravery, skill, and curiosity. The story of pioneering navigators is one of relentless pursuit of knowledge, fueled by the desire to explore the unknown. These intrepid explorers, often at great personal risk, ventured into territories that were previously thought to be unreachable. Their voyages not only expanded geographical knowledge but also paved the way for cultural exchanges, trade, and the eventual globalization of human society.

Among the earliest known navigators were the Polynesians, who embarked on incredible voyages across the vast Pacific Ocean thousands of years ago. Navigating by the stars, ocean currents, and bird flight patterns, they settled islands spread across a vast expanse of ocean, from Hawaii to New Zealand. These voyages were remarkable not only for their distance but also for their precision. The Polynesians developed sophisticated navigational techniques, such as the use of stick charts that represented wave patterns and island positions, allowing them to traverse thousands of miles of open water with remarkable accuracy.

In the Mediterranean, the Phoenicians emerged as master navigators around 1500 BCE. Their seafaring skills and extensive trade networks made them the dominant maritime power of their time. The Phoenicians are credited with circumnavigating Africa as early as the 6th century BCE, a feat that was not replicated for nearly two thousand years. Their expertise in shipbuilding and navigation laid the groundwork for future explorers and established important sea routes that facilitated trade and cultural exchange.

The Age of Exploration, which began in the 15th century, was a period of unprecedented maritime exploration. European nations, driven by the desire for new trade routes and territorial expansion, sponsored numerous expeditions. One of the most famous pioneers of this era was Christopher Columbus, an Italian navigator whose 1492 voyage across the Atlantic Ocean led to the European discovery of the Americas. Columbus's expedition, funded by the Spanish monarchy, was based on the belief that a westward sea route to Asia existed. Although he never reached Asia, his voyages opened up the New World to European exploration and colonization.

Another significant figure of the Age of Exploration was Ferdinand Magellan, a Portuguese navigator whose expedition became the first to circumnavigate the globe. In 1519, Magellan set sail with a fleet of five ships, aiming to find a westward route to the Spice Islands. Despite facing mutinies, treacherous waters, and his eventual death in the Philippines, Magellan's expedition successfully completed the journey in 1522 under the command of Juan Sebastián Elcano. This monumental voyage proved that the world was indeed round and that it was possible to sail around it.

The Portuguese also played a crucial role in pioneering navigation during this period. Vasco da Gama's voyage to India in 1498 opened up a sea route to the lucrative spice markets of Asia, bypassing the overland routes controlled by Middle Eastern and North African powers. This not only established Portugal as a dominant maritime power but also marked the beginning of European colonialism in Asia. Da Gama's successful navigation of the Cape of Good Hope demonstrated the potential for sea routes to connect distant parts of the world, leading to an era of global trade and exploration.

In the 18th century, British navigator James Cook embarked on a series of expeditions that significantly expanded the known world. Cook's voyages to the Pacific Ocean, including the exploration of Australia, New Zealand, and Hawaii, mapped vast stretches of previously uncharted territory. His meticulous charting and scientific observations contributed to the fields of geography, astronomy, and natural history. Cook's expeditions were notable not only for their discoveries but also for their relatively humane treatment of indigenous peoples, setting a standard for future explorers.

While European navigators often receive the most attention, it's important to recognize the contributions of other cultures to the field of navigation. Chinese admiral Zheng He, during the early 15th century, led seven major expeditions that reached as far as the east coast of Africa. Zheng He's fleet, consisting of massive treasure ships, was a testament to the advanced shipbuilding and navigational capabilities of the Ming Dynasty. These voyages established Chinese presence and influence across the Indian Ocean, facilitating trade and cultural exchanges long before European exploration began in earnest.

The advent of modern navigation owes much to these pioneering navigators and their voyages. Their achievements were made possible by advancements in nautical technology and navigational instruments. The development of the magnetic compass, astrolabe, and later the sextant enabled more precise navigation by allowing sailors to determine their position relative to the stars and magnetic north. These tools, combined with detailed maps and sea charts, provided the foundation for the exploration and mapping of the world.

Furthermore, the establishment of observatories and the publication of navigational manuals, such as the "Ephemerides" by German mathematician and astronomer Johannes Werner, played a crucial role in advancing navigational knowledge. These resources provided sailors with the information needed to calculate latitude and longitude, essential for accurate navigation.

The legacy of pioneering navigators is evident in the modern world. Their voyages not only expanded the geographical boundaries of the known world but also facilitated the exchange of ideas, cultures, and goods. This era of exploration laid the groundwork for the interconnected global society we live in today, where travel and communication across vast distances are commonplace.

In practical terms, the lessons learned from these early navigators continue to influence contemporary navigation. The principles of celestial navigation, the use of the compass, and the importance of accurate charting remain relevant, even in the age of GPS and satellite technology. Modern sailors and navigators still benefit from the knowledge and techniques developed by their predecessors.

For beginners interested in the field of navigation, understanding the history and contributions of pioneering navigators provides valuable insights into the evolution of this discipline. It highlights the importance of curiosity, innovation, and perseverance in overcoming the challenges of the unknown. By studying the achievements and methods of these early explorers, one can appreciate the complexity and skill involved in navigating the world's oceans and develop a deeper respect for the art and science of navigation.

Maritime Discoveries

The vastness and mystery of the world's oceans have always beckoned the adventurous spirit of humanity. Since ancient times, the desire to explore and understand the uncharted territories of the sea has driven countless maritime discoveries. These voyages have not only expanded our geographical knowledge but have also led to significant cultural exchanges, technological advancements, and economic growth.

The earliest known maritime discoveries date back to ancient civilizations such as the Egyptians, Phoenicians, and Greeks. These seafaring peoples ventured into the Mediterranean Sea, the Red Sea, and beyond, establishing trade routes and colonies. The Egyptians, for instance, conducted expeditions to Punt, a region believed to be located along the coast of East Africa or the Arabian Peninsula. These journeys, which took place around 2500 BCE, brought back valuable commodities such as gold, incense, and exotic animals, fueling the wealth and culture of ancient Egypt.

The Phoenicians, renowned for their shipbuilding and navigational skills, were instrumental in the exploration of the Mediterranean and beyond. By around 1200 BCE, they had established a network of colonies and trade routes stretching from the Levant to the Iberian Peninsula and North Africa. Their voyages extended into the Atlantic Ocean, where they are believed to have explored the coastlines of present-day Morocco and possibly even ventured as far as the British Isles. The Phoenicians' expertise in navigation and their development of the first alphabet significantly influenced subsequent maritime cultures.

In the classical era, the Greeks and Romans further expanded maritime exploration. Greek explorers such as Pytheas of Massalia sailed to the British Isles and possibly even reached Iceland around 300 BCE. Pytheas's accounts of the northern seas, including descriptions of the midnight sun and polar ice, provided valuable information to the ancient world.

The Romans, with their vast empire and advanced naval technology, conducted extensive maritime expeditions throughout the Mediterranean, the Black Sea, and the Atlantic coasts of Europe and Africa. Roman trade routes connected distant regions, facilitating the exchange of goods, ideas, and cultures.

The medieval period saw a resurgence of maritime exploration, particularly among the Vikings and the Islamic world. The Vikings, originating from Scandinavia, embarked on daring voyages across the North Atlantic, reaching as far as Greenland and North America. Around the year 1000 CE, Leif Erikson, a Norse explorer, established a settlement at Vinland, believed to be located in present-day Newfoundland, Canada. This remarkable achievement predates Christopher Columbus's arrival in the Americas by nearly 500 years, highlighting the Vikings' advanced seafaring capabilities.

Meanwhile, Islamic navigators and traders played a crucial role in maritime discoveries across the Indian Ocean and beyond. Muslim merchants and explorers, such as Ahmad ibn Fadlan and Ibn Battuta, documented their travels to regions including East Africa, India, Southeast Asia, and China. Their detailed accounts provided valuable insights into the cultures, geography, and trade networks of these distant lands. The establishment of maritime trade routes between the Islamic world and the Far East facilitated the exchange of goods, knowledge, and technologies, contributing to the flourishing of the medieval global economy.

The Age of Exploration, spanning the 15th to the 17th centuries, marked a pivotal era in maritime discoveries. European nations, driven by the desire for wealth, power, and knowledge, embarked on ambitious voyages of exploration. One of the most significant figures of this period was Christopher Columbus, whose 1492 voyage across the Atlantic Ocean led to the European discovery of the Americas. Sponsored by the Spanish monarchy, Columbus's expeditions opened up a new world of opportunities for European colonization, trade, and cultural exchange.

Another notable explorer of the Age of Exploration was Vasco da Gama, a Portuguese navigator who successfully reached India by sailing around the Cape of Good Hope in 1498. Da Gama's voyage established a direct sea route between Europe and Asia, bypassing the overland routes controlled by Middle Eastern and North African powers. This breakthrough not only secured Portugal's dominance in the spice trade but also paved the way for further maritime explorations and the establishment of European colonies in Asia.

Ferdinand Magellan, a Portuguese explorer sailing under the Spanish flag, achieved one of the most remarkable maritime discoveries by leading the first expedition to circumnavigate the globe. Departing in 1519, Magellan's fleet faced immense challenges, including treacherous waters, mutinies, and his own death in the Philippines. Despite these hardships, the expedition, under the command of Juan Sebastián Elcano, successfully completed the circumnavigation in 1522. This monumental achievement demonstrated the interconnectedness of the world's oceans and provided invaluable information about global geography.

The discoveries of the Age of Exploration were not limited to the Atlantic and Indian Oceans. European explorers also ventured into the Pacific Ocean, uncovering vast new territories and establishing trade routes. Spanish explorer Vasco Núñez de Balboa's crossing of the Isthmus of Panama in 1513 led to the discovery of the Pacific Ocean, while Portuguese navigator Ferdinand Magellan's expedition revealed the vast expanse of the Pacific and its numerous islands. These voyages significantly expanded European knowledge of the world and laid the groundwork for future explorations.

While European explorers often receive the most recognition, it is essential to acknowledge the contributions of other cultures to maritime discoveries. Chinese admiral Zheng He, during the early 15th century, led a series of seven major expeditions that reached as far as the east coast of Africa. Commanding a fleet of massive treasure ships, Zheng He's voyages established Chinese presence and influence across the Indian Ocean, promoting trade and cultural exchanges long before European exploration began in earnest.

The legacy of maritime discoveries extends beyond the geographical and economic realms. These voyages led to the exchange of ideas, technologies, and cultures, shaping the course of human history. The introduction of new crops, animals, and goods transformed societies and economies around the world. For example, the Columbian Exchange, resulting from Columbus's voyages, brought crops such as potatoes, maize, and tomatoes to Europe, while introducing Old World animals like horses and cattle to the Americas. This exchange had profound impacts on agriculture, cuisine, and daily life in both hemispheres.

Maritime discoveries also spurred advancements in navigation and shipbuilding. The development of navigational instruments such as the astrolabe, quadrant, and later the sextant enabled sailors to determine their position accurately. The creation of detailed sea charts and maps improved the safety and efficiency of maritime travel. Innovations in ship design, including the use of multiple masts and improved hull construction, allowed vessels to withstand longer and more challenging voyages. These technological advancements laid the foundation for the modern era of global trade and exploration.

For beginners interested in the field of maritime exploration, understanding the history and significance of these discoveries provides valuable insights into the evolution of navigation and global interconnectedness. The stories of these pioneering explorers highlight the importance of curiosity, innovation, and perseverance in overcoming the challenges of the unknown. By studying their achievements, one can appreciate the complexity and skill involved in navigating the world's oceans and develop a deeper respect for the art and science of maritime exploration.

The Legacy of Early Navigators

Navigating the vast unknown seas, early navigators laid the groundwork for modern seafaring and exploration. Their voyages, marked by bravery, innovation, and curiosity, transformed our understanding of the world and left an indelible legacy. These pioneers, hailing from different parts of the globe, not only mapped new territories but also connected distant cultures, fostering exchanges that shaped human history.

The Polynesians were among the earliest and most remarkable navigators, mastering the art of voyaging across the Pacific Ocean. Long before European explorers took to the seas, Polynesian navigators used sophisticated techniques to travel thousands of miles between islands. They relied on their acute knowledge of the stars, ocean currents, wind patterns, and bird migrations. This expertise enabled them to settle a vast expanse of the Pacific, from Hawaii to New Zealand to Easter Island, creating a network of interconnected cultures that thrived on trade and shared knowledge.

In the Indian Ocean, Arab and Persian sailors played a crucial role in early navigation. From as early as the 9th century, they sailed from the Arabian Peninsula to East Africa, India, and Southeast Asia. Using the monsoon winds, these mariners developed reliable trade routes that facilitated the exchange of goods such as spices, textiles, and precious stones. Their navigational skills were enhanced by the use of the astrolabe, a device that allowed them to measure the altitude of stars and calculate their position at sea. This innovation, along with their detailed maps and charts, laid the groundwork for future explorers.

The Vikings, fearless seafarers from Scandinavia, expanded their reach across the North Atlantic during the 8th to 11th centuries. Their longships, designed for speed and versatility, enabled them to explore and establish settlements in Iceland, Greenland, and even North America, long before Columbus. Notably, Leif Erikson's voyage around the year 1000 AD led to the Norse settlement at Vinland, believed to be in present-day Newfoundland. The Vikings' navigation relied heavily on coastal landmarks, the sun's position, and an intimate understanding of the natural environment, showcasing their resourcefulness and adaptability.

During the Age of Exploration in the 15th and 16th centuries, European navigators embarked on ambitious voyages that would reshape the world. Portuguese explorers, under the patronage of Prince Henry the Navigator, pioneered the exploration of the African coast. Bartolomeu Dias's rounding of the Cape of Good Hope in 1488 and Vasco da Gama's subsequent voyage to India in 1498 opened new maritime routes to Asia, challenging the overland trade monopolies of the time. These expeditions were made possible by advancements in ship design, such as the caravel, and navigational tools like the quadrant and compass.

Christopher Columbus, an Italian navigator sponsored by Spain, is often credited with "discovering" the New World in 1492. His transatlantic voyages, though initially seeking a westward route to Asia, led to the European awareness of the American continents. Columbus's journeys marked the beginning of sustained contact between Europe and the Americas, resulting in profound cultural exchanges and the eventual colonization of vast territories. His voyages were navigational feats that relied on dead reckoning and celestial navigation, showcasing the evolving maritime skills of the period.

Ferdinand Magellan, a Portuguese explorer in the service of Spain, achieved the first circumnavigation of the globe, a monumental milestone in navigation history. Departing in 1519, Magellan's expedition faced numerous challenges, including treacherous seas, mutinies, and Magellan's own death in the Philippines. Despite these obstacles, the expedition, led to completion by Juan Sebastián Elcano, returned to Spain in 1522, proving that the Earth was indeed round and interconnected by oceans. This voyage provided invaluable data on global geography and ocean currents, solidifying the importance of maritime exploration.

The contributions of non-European navigators are equally significant. Zheng He, a Chinese admiral of the Ming Dynasty, led seven major expeditions between 1405 and 1433. Commanding a fleet of enormous treasure ships, Zheng He's voyages reached as far as the east coast of Africa, the Arabian Peninsula, and Southeast Asia. These expeditions were not only demonstrations of Chinese naval power but also facilitated trade, diplomacy, and cultural exchange. Zheng He's navigational achievements were supported by advanced shipbuilding techniques and the use of the magnetic compass, an invention that would later revolutionize global navigation.

The legacy of early navigators extends beyond their geographical discoveries. Their voyages fostered the exchange of goods, ideas, technologies, and cultures, creating a more interconnected world. The Columbian

Exchange, for instance, introduced new crops and livestock to different continents, profoundly affecting agriculture and diets. The spread of navigational knowledge and tools, such as the compass and astrolabe, revolutionized sea travel, making long-distance voyages safer and more efficient.

Moreover, these early navigators inspired a spirit of exploration and curiosity that continues to drive human endeavors. Their stories of bravery and perseverance in the face of the unknown serve as timeless reminders of the potential for discovery and innovation. The maps and charts they created, often at great personal risk, laid the foundations for modern cartography and navigation, enabling future generations to build upon their achievements.

For beginners interested in the history of navigation, understanding the legacy of these early navigators offers valuable lessons. Their voyages illustrate the importance of preparation, adaptability, and a deep understanding of the natural world. Whether it was the Polynesians reading the stars, the Vikings interpreting coastal features, or the Europeans using celestial navigation, each culture developed unique methods to master the seas. Studying their techniques and innovations can provide insights into the evolution of navigation and the enduring human quest for exploration.

In practical terms, aspiring navigators can learn from the meticulous planning and resourcefulness of these pioneers. The construction of reliable ships, the development of accurate maps, and the use of navigational instruments were all critical to their success. Today's navigators can draw parallels to modern practices, where technology and traditional knowledge combine to ensure safe and efficient maritime travel. The legacy of early navigators underscores the importance of continuous learning and adaptation in the face of new challenges.

The spirit of these early explorers is alive in contemporary maritime practices. Modern navigators, whether in commercial shipping, scientific research, or recreational sailing, benefit from the advancements and discoveries made by their predecessors. Satellite navigation, advanced weather forecasting, and sophisticated ship design are all part of the legacy handed down through centuries of maritime exploration.

Chapter 4

Exploring the New World

Christopher Columbus and His Voyages

Christopher Columbus, a name that evokes images of grand ships and uncharted territories, remains one of history's most renowned explorers. Born in Genoa, Italy, around 1451, Columbus grew up in a seafaring environment, which naturally led him to a life on the seas. His early experiences as a trader and navigator laid the groundwork for his ambitious vision: to find a westward route to Asia. This vision, driven by the desire for wealth, fame, and the spread of Christianity, would eventually lead to his historic voyages across the Atlantic and the opening of the New World to European exploration and colonization.

Columbus's belief in a westward route to Asia was based on several misconceptions. He underestimated the Earth's circumference and overestimated the size of Asia, believing that Japan lay approximately 2,400 miles west of Europe, rather than the actual 12,000 miles. Despite these errors, Columbus was convinced that such a route would be shorter and more profitable than the perilous overland journey through the Middle East or the lengthy voyage around Africa. His conviction led him to seek patronage for his expedition. After years of lobbying European monarchs, Columbus finally secured the support of King Ferdinand and Queen Isabella of Spain in 1492.

On August 3, 1492, Columbus set sail from the port of Palos with three ships: the Niña, the Pinta, and the Santa María. His crew comprised experienced sailors and navigators, many of whom were skeptical or fearful of the unknown. The journey across the Atlantic was fraught with challenges, including unpredictable weather, the threat of mutiny, and the constant fear of running out of supplies. Columbus maintained his crew's morale through a combination of strict discipline, charismatic leadership, and promises of wealth and glory upon their return.

After about two months at sea, on October 12, 1492, land was finally sighted. Columbus and his crew had reached an island in the Bahamas, which he named San Salvador. Believing he had found islands off the coast of Asia, Columbus referred to the inhabitants as "Indians." This misidentification would persist for centuries, shaping European perceptions of the New World. The encounter was marked by initial curiosity and cautious interactions. Columbus's journals record both admiration for the natives' hospitality and an eagerness to convert them to Christianity.

Columbus explored the Caribbean for several months, visiting present-day Cuba and Hispaniola (now Haiti and the Dominican Republic). He established a small settlement, La Navidad, on Hispaniola, leaving behind a contingent of men to build a foothold for future expeditions. When Columbus returned to Spain in March 1493, he brought with him tangible proof of his findings: gold, exotic plants, animals, and a few native inhabitants. His return was celebrated as a monumental achievement, and he was hailed as the Admiral of the Ocean Sea.

Encouraged by his initial success, Columbus embarked on three more voyages between 1493 and 1504. Each expedition aimed to further explore and exploit the newfound territories. On his second voyage, Columbus returned with a larger fleet and more settlers, seeking to establish permanent colonies. However, he faced mounting challenges, including resistance from indigenous populations, internal strife among settlers, and difficulties in managing the colonies. Despite these setbacks, Columbus continued to explore the Caribbean, discovering new islands and mapping their coastlines.

Columbus's third voyage in 1498 took him farther south, where he encountered the mainland of South America at the Orinoco River delta. This discovery hinted at the vastness of the New World, but Columbus remained convinced that he was exploring the outskirts of Asia. His tenure as governor of the colonies, however, was marred by accusations of mismanagement and brutality. Reports of harsh treatment of both settlers and natives led to his arrest and return to Spain in chains in 1500. Although he was eventually released and his titles restored, his reputation had been significantly tarnished.

His final voyage in 1502 was a desperate attempt to find a passage to Asia. Columbus explored the coast of Central America, from Honduras to Panama, but was thwarted by adverse weather and hostile encounters with indigenous groups. Stranded for nearly a year in Jamaica, he and his men endured severe hardships before being rescued and returning to Spain in 1504. Columbus's health deteriorated, and he spent his final years seeking recognition and compensation for his contributions. He died in 1506, still believing he had found a new route to Asia.

The legacy of Christopher Columbus is complex and multifaceted. On one hand, his voyages initiated an era of exploration and colonization that transformed the world. The Columbian Exchange, the widespread transfer of plants, animals, culture, human populations, technology, and ideas between the Americas and the Old World, had profound and lasting impacts. Crops like potatoes, tomatoes, and maize became staples in Europe, while Old World animals such as horses and cattle transformed the Americas. The exchange also brought devastating diseases like smallpox to the indigenous populations, leading to dramatic population declines.

On the other hand, Columbus's expeditions set the stage for centuries of exploitation, conquest, and colonization that had dire consequences for native populations. The encomienda system, established by the Spanish crown, granted colonists the right to extract labor and tribute from indigenous people, leading to widespread abuse and forced labor. The transatlantic slave trade, which began in earnest in the decades following Columbus's voyages, further compounded the suffering and displacement of countless individuals.

Despite these darker aspects, Columbus's achievements in navigation and exploration cannot be overlooked. His voyages demonstrated the feasibility of transatlantic travel and opened the way for future explorers like Amerigo Vespucci, who first recognized the Americas as a distinct continent, and Ferdinand Magellan, whose expedition completed the first circumnavigation of the globe. The maps and navigational data collected during these early voyages contributed significantly to the growing body of knowledge about the world's geography.

For beginners studying the history of exploration, Columbus's story offers valuable lessons in perseverance, ambition, and the complexities of historical legacy. His determination to pursue a bold vision, despite numerous setbacks and misunderstandings, underscores the importance of resilience and innovation. At the same time, the consequences of his voyages remind us of the ethical considerations and human costs that accompany exploration and discovery.

In practical terms, understanding Columbus's voyages provides a foundation for comprehending the larger narrative of European exploration and colonization. His navigational techniques, use of available technology, and interactions with indigenous peoples offer insights into the challenges and dynamics of early maritime expeditions. As we reflect on Columbus's legacy, it is crucial to recognize both his contributions to global knowledge and the profound impacts of his voyages on the world's history and cultures.

The Conquests of Hernán Cortés

Hernán Cortés, born in 1485 in Medellín, Spain, emerged as one of the most notable figures in the era of Spanish exploration and conquest. His journey to fame and infamy began when he sailed from Spain to the New World in 1504, driven by the lure of adventure and riches. Cortés's ambitions and strategic acumen would eventually lead to the fall of the mighty Aztec Empire, reshaping the course of history in the Americas.

Cortés first arrived in Hispaniola, where he quickly distinguished himself as a capable and ambitious man. His talents did not go unnoticed, and he soon joined Diego Velázquez's expedition to conquer Cuba in 1511. As a reward for his services, Velázquez granted Cortés a large estate and a number of indigenous laborers. Despite this success, Cortés's restless spirit and insatiable ambition pushed him to seek greater opportunities. By 1518, he had convinced Velázquez to appoint him as the leader of an expedition to explore and secure the mainland of Mexico.

Setting sail in February 1519 with a fleet of eleven ships, around 600 men, and a few horses, Cortés embarked on an expedition that would become legendary. Upon reaching the coast of present-day Mexico, he founded the settlement of Veracruz. This move was strategic; by establishing a colony directly under the authority of the Spanish crown, he circumvented Velázquez's control. His leadership and assertiveness were evident as he quickly organized his men and began forging alliances with local indigenous groups who were hostile to the dominant Aztec Empire.

Cortés's encounter with the Aztecs began with a mix of diplomacy and intimidation. He learned of the Aztec capital, Tenochtitlán, and its emperor, Moctezuma II, from local informants. The wealth and grandeur of the Aztec Empire tantalized Cortés and his men. Utilizing a combination of force, negotiation, and the strategic use of interpreters like Malintzin (also known as La Malinche), who played a crucial role in communication and alliance-building, Cortés made his way inland.

Moctezuma, aware of the Spaniards' arrival and uncertain of their intentions, opted to welcome Cortés diplomatically. He perhaps saw the Spaniards as potential allies or even divine beings, as some accounts suggest that the Aztecs initially believed Cortés might be the god Quetzalcoatl returning. In November 1519, Cortés and his men were received in Tenochtitlán with great ceremony. The city, built on a series of islands in Lake Texcoco, was a marvel of engineering and culture, unlike anything the Spaniards had ever seen.

Despite the warm reception, tensions quickly escalated. Cortés, ever the strategist, took Moctezuma hostage in his own palace, hoping to control the empire through the emperor. This bold move temporarily consolidated Cortés's power but also sowed the seeds of rebellion. The situation deteriorated further when Cortés had to leave Tenochtitlán to confront a punitive expedition sent by Velázquez to arrest him. Leaving his lieutenant Pedro de Alvarado in charge, Cortés marched to the coast, defeated the Velázquez-led force, and persuaded many of the soldiers to join his cause.

Upon his return to Tenochtitlán, Cortés found the city in turmoil. Alvarado's heavy-handed tactics had provoked the Aztecs, leading to a full-scale uprising. In the ensuing chaos, Moctezuma was killed— accounts vary on whether he was murdered by the

Spaniards or killed by his own people. The Spaniards, now besieged in the palace, faced a desperate situation. In the dramatic and brutal retreat known as La Noche Triste (The Night of Sorrows), Cortés and his men fought their way out of the city, suffering significant losses as they fled across a causeway under relentless Aztec attack.

Regrouping in Tlaxcala, an indigenous state that was a fierce rival of the Aztecs and had allied with the Spaniards, Cortés prepared for a final assault on Tenochtitlán. He built a fleet of brigantines to control the lake and cut off the city's supplies. In May 1521, after months of preparation and skirmishes, Cortés launched the siege of Tenochtitlán. The battle was fierce and protracted, with both sides suffering heavy casualties. The Spaniards, with their superior weaponry and tactics, slowly gained the upper hand. Finally, on August 13, 1521, the Aztec capital fell, marking the end of the Aztec Empire.

The conquest of Tenochtitlán was not merely a military victory but a turning point with profound implications. Cortés established Mexico City on the ruins of the Aztec capital, which became the center of Spanish colonial administration in the New World. His conquests opened the floodgates for further Spanish exploration and colonization, leading to the vast expansion of the Spanish Empire in the Americas.

However, Cortés's legacy is deeply controversial. While he is often celebrated for his daring and strategic genius, his actions also brought immense suffering. The conquest led to the decimation of the indigenous population through warfare, enslavement, and diseases such as smallpox, which the Europeans unwittingly introduced. The cultural and social fabric of Mesoamerican civilizations was irrevocably altered as Spanish rule imposed new systems of governance, religion, and economy.

Cortés himself faced numerous challenges following his conquests. His relationship with the Spanish crown became strained as his rivals at court sought to undermine him. Despite his contributions, he struggled to gain the recognition and rewards he believed he deserved. In 1528, he returned to Spain to defend his actions and seek favor at the royal court. Although King Charles I received him, Cortés was never fully reinstated to his former position of power. He returned to Mexico in 1530, where he continued to explore and govern his estates but faced ongoing legal battles and diminishing influence.

In his later years, Cortés participated in several expeditions, including a venture to explore the northern coasts of Mexico and the Gulf of California. Despite his persistent efforts to recapture his earlier glory, he spent his final years largely marginalized from the centers of power. Hernán Cortés died in Spain in 1547, leaving behind a complex and contested legacy.

For those studying the era of conquest and colonization, Cortés's story offers a compelling narrative of ambition, conflict, and transformation. His actions illustrate the interplay of personal ambition and broader historical forces, shedding light on the dynamics of power, culture, and resistance. Understanding Cortés's conquests provides insights into the profound changes that shaped the early modern world and the enduring impacts on the societies that were forever altered by these encounters.

The Expeditions of Francisco Pizarro

Francisco Pizarro, born around 1478 in Trujillo, Spain, emerged from humble beginnings to become one of the most formidable conquistadors of the Spanish Empire. His expeditions into South America, particularly his conquest of the Inca Empire, were both daring and transformative, altering the course of history and expanding the reach of Spanish influence.

Pizarro's early life was marked by hardship and a lack of formal education. The illegitimate son of Gonzalo Pizarro, a colonel of infantry, young Francisco spent much of his youth tending pigs. Yet, his ambitions stretched far beyond the fields of Extremadura. Drawn by tales of wealth and adventure in the New World, he sailed for Hispaniola in 1502, where he began his career as an explorer and soldier.

His early years in the New World were spent in various roles, including participating in Vasco Núñez de Balboa's expedition that discovered the Pacific Ocean in 1513. This journey across the Isthmus of Panama ignited Pizarro's desire for exploration. He heard rumors of a rich and powerful empire to the south, a land filled with gold and silver, which fueled his ambitions.

Pizarro's first major expedition southward began in 1524, in partnership with Diego de Almagro and

Hernando de Luque. This initial venture was fraught with difficulties, including harsh weather, disease, and resistance from indigenous peoples. The expedition reached as far as the present-day Colombia-Ecuador border but returned to Panama with little to show for their efforts. Undeterred, Pizarro and his partners launched a second expedition in 1526, which also faced significant challenges but provided crucial intelligence about the Inca Empire.

It was during this second expedition that Pizarro's persistence began to pay off. After enduring severe hardships, including a perilous winter on Gallo Island, Pizarro and his men finally reached the northern territories of the Inca Empire. They encountered an advanced civilization with impressive architecture and a complex social structure, confirming the tantalizing rumors of immense wealth. Pizarro sent a delegation back to Panama to secure more resources and reinforcements for a full-scale conquest.

In 1528, Pizarro returned to Spain to seek royal support for his ambitious plans. King Charles I granted him the authority to conquer and govern the newly discovered lands, providing him with the necessary legitimacy and resources for his enterprise. Armed with this royal endorsement, Pizarro embarked on his third and final expedition in 1531, accompanied by a small but determined force of around 180 men, 27 horses, and two cannons.

Upon landing on the northern coast of Peru, Pizarro's force moved inland, encountering various indigenous groups along the way. His strategy of forming alliances with discontented local tribes proved crucial, as many were eager to rebel against Inca rule. These alliances provided Pizarro with additional manpower and local knowledge, significantly bolstering his small contingent.

The Inca Empire, at this time, was weakened by a civil war between two brothers, Atahualpa and Huáscar, who were vying for the throne. Pizarro astutely exploited this internal strife. In November 1532, he arranged a meeting with Atahualpa in the town of Cajamarca. Despite their numerical disadvantage, Pizarro and his men launched a surprise attack during the meeting, capturing Atahualpa and killing thousands of Inca warriors in the process.

The capture of Atahualpa was a turning point. Pizarro held the emperor for ransom, demanding an enormous sum of gold and silver, which the Incas delivered, filling an entire room with precious metals. Despite receiving the ransom, Pizarro executed Atahualpa in 1533, fearing that leaving him alive would perpetuate resistance. This act plunged the Inca Empire into chaos, facilitating its conquest.

With Atahualpa's death, Pizarro marched towards the Inca capital of Cuzco, encountering sporadic resistance along the way. He finally entered the city in November 1533, marking the culmination of his conquest. The fall of Cuzco effectively dismantled the Inca state, although sporadic resistance and rebellions continued for several years.

Pizarro established a new capital, Lima, in 1535, which became the administrative and political center of Spanish Peru. His governance, however, was marred by conflicts with his former partner, Diego de Almagro. The two conquistadors had a falling out over

territorial disputes, culminating in a bloody conflict
known as the Battle of Las Salinas in 1538, where
Pizarro's forces defeated and executed Almagro.

Despite his military successes, Pizarro's later years
were overshadowed by political intrigue and betrayal.
He faced ongoing challenges from Almagro's
supporters and other Spanish rivals. In 1541, a group
of Almagro's loyalists assassinated Pizarro in his
palace in Lima, ending his tumultuous and
controversial life.

Francisco Pizarro's expeditions and conquests had
far-reaching consequences. The incorporation of the
vast and wealthy Inca Empire into the Spanish realm
significantly boosted Spain's economy, flooding
Europe with gold and silver. The conquest also
marked the beginning of a period of profound social
and cultural upheaval in the Andes. The Spanish
imposed their language, religion, and administrative
systems, leading to the gradual erosion of indigenous
traditions and institutions.

The demographic impact was equally devastating. The
introduction of European diseases, to which the
indigenous population had no immunity, caused
widespread mortality. Estimates suggest that the
indigenous population of the Andes declined by as
much as 90% within the first century of Spanish rule.

Pizarro's legacy is complex and contested. He is often
portrayed as a ruthless conqueror whose actions
brought immense suffering to the indigenous peoples
of South America. Yet, his audacity, strategic prowess,
and ability to navigate the treacherous political
landscape of the New World were undeniable. His

conquests opened the door for further Spanish exploration and colonization, profoundly shaping the history of the Americas.

For contemporary readers, Pizarro's story offers valuable insights into the dynamics of conquest and colonization. It highlights the interplay of ambition, greed, and strategic calculation that drove European expansion into the New World. It also underscores the resilience and adaptability of indigenous societies in the face of overwhelming external pressures.

The Impact on Indigenous Cultures

The arrival of European explorers in the Americas marked the beginning of a profound and often devastating transformation for indigenous cultures. The impact on these societies varied widely, but common themes of disruption, adaptation, and resistance emerged across different regions. The encounter between indigenous peoples and European colonizers resulted in significant changes in demographics, social structures, economies, and cultural practices.

The introduction of European diseases had a catastrophic effect on indigenous populations. Smallpox, influenza, measles, and other diseases spread rapidly and with deadly efficiency, decimating communities that had no natural immunity. Some estimates suggest that the indigenous population of the Americas declined by as much as 90% within the first century of contact. In many cases, entire villages and even larger political entities were wiped out, leading to significant social and political upheaval.

The demographic collapse had immediate and far-reaching consequences. With a dramatic reduction in population, the labor force necessary to sustain agricultural and economic activities was severely diminished. This, in turn, led to food shortages and further weakened indigenous societies. The loss of leaders and skilled artisans disrupted governance and cultural transmission, compounding the sense of chaos and loss.

European colonization also brought about significant changes in land use and ownership. Europeans introduced new agricultural practices and crops, such as wheat, barley, and livestock, which often conflicted with traditional land management systems. The imposition of European concepts of private property and legal systems further disrupted indigenous land tenure, leading to the dispossession of lands that had been held communally for generations. This loss of land not only had economic implications but also struck at the heart of many indigenous cultures, which were deeply connected to their ancestral territories.

The Spanish encomienda system, in particular, had a profound impact on indigenous societies in Latin America. Under this system, Spanish settlers were granted the right to extract labor and tribute from indigenous communities in exchange for providing protection and religious instruction. In practice, this often led to severe exploitation and abuse. Encomenderos forced indigenous people to work in mines, plantations, and construction projects under harsh conditions, leading to further population decline and social disintegration.

The cultural impact of European colonization was equally profound. The Catholic Church played a central role in the Spanish colonization process, aiming to convert indigenous peoples to Christianity. Missionaries established schools and missions to educate and convert the indigenous population, often using coercive methods. Indigenous religious practices were suppressed, and sacred sites were destroyed or repurposed for Christian worship. This led to a significant loss of traditional knowledge and spiritual practices, although in many cases, indigenous peoples found ways to syncretize their beliefs with Christianity, creating unique cultural blends that persist to this day.

Language was another area where the impact of colonization was deeply felt. Spanish, Portuguese, and later other European languages became the dominant languages of administration, education, and trade. Indigenous languages were often marginalized and faced decline as younger generations were educated in European languages. This linguistic shift contributed to the erosion of cultural heritage and identity, as language is a key carrier of cultural knowledge and traditions.

Despite these profound challenges, indigenous cultures demonstrated remarkable resilience and adaptability. Many communities found ways to preserve and adapt their cultural practices in the face of European domination. Some indigenous leaders engaged with European legal and political systems to defend their communities' rights and negotiate better terms of coexistence. Others maintained their cultural practices in secret or integrated new elements into their traditions to ensure their survival.

The mestizaje, or mixing of indigenous and European peoples, also played a significant role in shaping the cultural landscape of the Americas. This process of cultural and biological blending resulted in the emergence of new, hybrid identities and cultures. Mestizo populations, which combined elements of both indigenous and European heritage, became a significant demographic group in many parts of Latin America. This blending of cultures also influenced art, music, cuisine, and other cultural expressions, creating rich and diverse cultural traditions.

Resistance to European domination took many forms, from armed uprisings to everyday acts of defiance. Indigenous leaders such as Túpac Amaru II in Peru and Popé in New Mexico led significant rebellions against Spanish rule. These uprisings, although often brutally suppressed, demonstrated the enduring spirit and agency of indigenous peoples. In other cases, resistance manifested in more subtle ways, such as the preservation of traditional practices, the use of indigenous languages, and the transmission of cultural knowledge through oral traditions.

The impact of European colonization on indigenous cultures is a complex and multifaceted story, marked by both profound loss and enduring resilience. While the initial period of contact brought about significant disruption and transformation, indigenous cultures have continued to evolve and adapt in response to changing circumstances. Today, indigenous peoples across the Americas continue to assert their rights, preserve their cultural heritage, and contribute to the rich tapestry of global cultural diversity.

In contemporary times, there has been a growing recognition of the importance of preserving and revitalizing indigenous cultures. Efforts to document and promote indigenous languages, protect sacred sites, and support traditional cultural practices are gaining momentum. Indigenous leaders and communities are actively engaged in these efforts, often in collaboration with governments, non-governmental organizations, and international bodies.

One significant development has been the increasing visibility and influence of indigenous movements on the global stage. Indigenous activists and organizations have played a key role in advocating for indigenous rights, environmental protection, and social justice. The United Nations Declaration on the Rights of Indigenous Peoples, adopted in 2007, represents a milestone in the recognition of indigenous peoples' rights to self-determination, land, and cultural preservation.

Education has also emerged as a critical area for supporting indigenous cultural revitalization. Schools and educational programs that incorporate indigenous languages, knowledge systems, and cultural practices are helping to foster a sense of pride and identity among indigenous youth. These efforts are not only important for cultural preservation but also for addressing the social and economic challenges faced by many indigenous communities.

The Columbian Exchange

When Christopher Columbus made his first voyage to the New World in 1492, he unknowingly initiated a massive ecological and cultural exchange between the Americas, Europe, Africa, and Asia. This process, known as the Columbian Exchange, profoundly altered the world's ecosystems, economies, societies, and even diets. The movement of plants, animals, diseases, and people across continents reshaped the course of history in ways that remain evident today.

One of the most significant aspects of the Columbian Exchange was the introduction of new crops to different parts of the world. From the Americas, crops such as maize, potatoes, tomatoes, and cassava were transported to Europe, Africa, and Asia. These crops often thrived in their new environments and became staples of diets far from their places of origin. For instance, potatoes, originally from the Andes, became crucial in European agriculture, especially in Ireland and Russia. Similarly, maize spread rapidly across Africa and Asia, becoming a critical source of food and contributing to population growth.

Conversely, European settlers introduced Old World crops to the Americas, including wheat, barley, rice, and sugarcane. These crops often replaced traditional indigenous agriculture or were integrated into existing farming systems. The introduction of sugarcane, in particular, had far-reaching consequences. Sugar plantations, established first in the Caribbean and later in the Americas, fueled the demand for labor, leading to the transatlantic slave trade. The labor-intensive nature of sugar cultivation necessitated a massive influx of African slaves, which in turn had profound social and demographic impacts.

The exchange of animals was another transformative aspect of the Columbian Exchange. Europeans brought livestock such as horses, cattle, pigs, and sheep to the New World. The introduction of horses, in particular, revolutionized indigenous societies. Plains tribes like the Comanche and the Sioux adopted the horse for hunting, warfare, and transportation, dramatically altering their ways of life. Horses provided new mobility and power, enabling these tribes to expand their territories and enhance their hunting capabilities.

Cattle and pigs also had significant impacts. Cattle ranching became a major economic activity in parts of the Americas, from the pampas of Argentina to the plains of Texas. The proliferation of livestock often led to environmental changes, such as overgrazing and soil erosion, which affected indigenous agricultural practices. Pigs, known for their adaptability and rapid reproduction, often became feral and spread across landscapes, sometimes disrupting local ecosystems and competing with native species.

The exchange of diseases, however, was perhaps the most devastating consequence of the Columbian Exchange. Europeans brought with them pathogens such as smallpox, influenza, measles, and typhus, to which indigenous populations had no immunity. The effects were catastrophic, with mortality rates in some areas reaching as high as 90%. Entire communities were decimated, leading to significant social and political upheaval. The demographic collapse weakened resistance to European colonization and facilitated the conquest and settlement of the Americas.

Conversely, the movement of people also played a critical role in the Columbian Exchange. European settlers, African slaves, and later Asian laborers migrated to the Americas, bringing with them their cultures, languages, and traditions. This movement of people led to the creation of new, hybrid cultures. In Latin America, for instance, the blending of indigenous, African, and European influences gave rise to rich and diverse cultural expressions, from music and dance to cuisine and religion.

The Columbian Exchange also had profound economic impacts. The introduction of new crops and livestock boosted agricultural productivity and supported population growth. In Europe, the influx of silver and gold from the Americas led to increased wealth and the expansion of trade networks. However, this wealth was often concentrated in the hands of European elites, leading to increased social stratification and economic disparities.

The transatlantic slave trade, driven by the demand for labor on American plantations, had enduring economic consequences. The forced migration of millions of Africans to the Americas not only had a devastating impact on African societies but also laid the foundations for systemic racial inequalities that persisted for centuries. The labor of enslaved Africans was a key factor in the economic development of the Americas, particularly in the production of cash crops like sugar, tobacco, and cotton.

Environmental changes resulting from the Columbian Exchange were also significant. The introduction of new species led to ecological transformations, sometimes with unintended consequences. Invasive species, such as European rats and weeds, disrupted local ecosystems and displaced native species. The large-scale cultivation of cash crops often led to deforestation and soil depletion, altering landscapes and affecting biodiversity.

The Columbian Exchange also had cultural and culinary impacts that continue to shape our world today. The global movement of crops and foods led to the blending of cuisines and the creation of new culinary traditions. Imagine Italian cuisine without tomatoes, or Irish cuisine without potatoes. The introduction of chili peppers from the Americas transformed cuisines in India, China, and Southeast Asia, adding a new dimension of spice and flavor.

In Africa, the introduction of American crops like cassava and maize provided new food sources that could be cultivated in diverse environments, supporting population growth and agricultural diversity. Similarly, the spread of Old World crops like wheat and rice to the Americas introduced new dietary staples and agricultural practices.

The Columbian Exchange also facilitated the spread of knowledge and ideas. European explorers and settlers brought with them not only crops and animals but also technologies, scientific knowledge, and cultural practices. The exchange of agricultural techniques, medicinal practices, and navigational skills enriched societies on both sides of the Atlantic. Indigenous knowledge of plants and ecosystems, for instance, was often integrated into European scientific understanding, influencing fields such as botany and medicine.

Despite its many benefits, the Columbian Exchange also had profound ethical and moral implications. The exploitation of indigenous peoples and the enslavement of Africans were integral to the economic systems that emerged from this exchange. The legacy of these practices continues to affect contemporary societies, shaping debates around social justice, reparations, and historical memory.

The Columbian Exchange was a complex and multifaceted process that reshaped the world in profound ways. Its impacts were felt not only in the immediate aftermath of contact but continue to resonate in contemporary global systems. Understanding this historical phenomenon requires recognizing both its transformative benefits and its devastating costs. It is a testament to the interconnectedness of human societies and the ways in which cultural, biological, and economic exchanges can drive historical change.

Chapter 5

Conquering the Poles

The Quest for the North Pole

The allure of the North Pole has captivated explorers, scientists, and adventurers for centuries. This remote, icy region at the top of the world represents one of the last great frontiers of human exploration. The quest to reach the North Pole is a tale of ambition, endurance, and courage, intersecting with advancements in navigation, science, and technology.

In the early 19th century, the North Pole was largely a mystery. No one knew for certain what lay at the top of the world—whether it was a sea, a continent, or a vast expanse of ice. The quest began in earnest with the expeditions of British explorers like Sir John Ross and Sir William Parry, who sought to chart the Arctic and find the elusive Northwest Passage. Their voyages, fraught with peril and hardship, laid the groundwork for future exploration, even as they fell short of reaching the Pole itself.

One of the earliest and most famous attempts was led by Sir John Franklin in 1845. Franklin's expedition, consisting of two ships, the HMS Erebus and the HMS Terror, aimed to chart the last uncharted section of the Northwest Passage. Tragically, Franklin and his crew disappeared, prompting numerous search missions that gradually revealed the harsh realities of Arctic exploration. The fate of Franklin's expedition remained a haunting mystery until the discovery of the ships' wrecks in the 21st century.

As the 19th century progressed, technological advancements and a growing body of geographical knowledge spurred new expeditions. In 1879, the American explorer George W. De Long led the ill-fated Jeannette expedition, which ended in disaster when the ship was trapped and crushed by ice. De Long and his crew struggled across the ice, but most perished in the attempt. Despite its tragic end, the Jeannette expedition provided valuable data on Arctic conditions and underscored the immense challenges of polar travel.

The dawn of the 20th century saw a renewed fervor for Arctic exploration, driven by national pride and scientific curiosity. The race to the North Pole became a highly publicized contest, with American and European explorers vying for the honor. Robert E. Peary, an American naval officer, emerged as a leading figure in this quest. After several expeditions, Peary claimed to have reached the North Pole on April 6, 1909, accompanied by his assistant Matthew Henson and four Inuit guides. Peary's claim was initially accepted but later came under scrutiny due to inconsistencies in his navigational records and the absence of independent verification.

Simultaneously, Frederick A. Cook, another American explorer, claimed to have reached the North Pole a year earlier, in 1908. Cook's assertion sparked controversy and debate, as it lacked corroborating evidence and was widely disputed by the scientific community. The rivalry between Peary and Cook highlighted the intense competition and the difficulties in verifying such monumental achievements in an era before satellite navigation and instant communication.

The quest for the North Pole was not solely the domain of Americans. Norwegian explorer Roald Amundsen, famous for his successful expedition to the South Pole, also turned his attention northward. In 1926, Amundsen, along with the Italian engineer Umberto Nobile and American aviator Lincoln Ellsworth, flew over the North Pole in the airship Norge. This marked the first verified flight over the Pole, demonstrating the potential of air travel for Arctic exploration.

The mid-20th century brought new approaches to polar exploration, driven by advances in aviation and icebreaker technology. In 1958, the United States submarine USS Nautilus became the first vessel to reach the North Pole, traveling beneath the ice cap. This historic voyage underscored the strategic and scientific importance of the Arctic during the Cold War, as both the United States and the Soviet Union sought to assert their presence in the region.

In 1968, American explorer Ralph Plaisted and his team became the first to reach the North Pole by surface travel using snowmobiles. This achievement, verified by satellite navigation, marked a significant milestone in Arctic exploration. Plaisted's journey demonstrated the feasibility of modern transportation methods in the harsh polar environment and opened the door for future expeditions by various means.

The late 20th century saw an increasing focus on scientific research in the Arctic. The International

Geophysical Year (1957-1958) and subsequent multinational collaborations aimed to study the Arctic's climate, geology, and ecosystems. These efforts highlighted the critical role of the Arctic in global climate systems and underscored the need for international cooperation in polar research.

Environmental concerns also began to shape the discourse around the North Pole. The effects of climate change became increasingly evident, with rising temperatures leading to the melting of sea ice and permafrost. These changes not only threatened the fragile Arctic ecosystem but also had global implications, such as rising sea levels and altered weather patterns. The quest for the North Pole thus evolved from a focus on conquest and adventure to a deeper understanding of the region's significance for the planet's health.

In recent decades, the North Pole has become a site for both scientific inquiry and adventure tourism. Modern explorers, equipped with advanced technology and support systems, continue to journey to the Pole, contributing to our knowledge of this remote region. Scientific missions study the impacts of climate change, monitor wildlife populations, and investigate the Arctic's unique geology and oceanography.

The quest for the North Pole is a testament to human curiosity, resilience, and the drive to explore the unknown. It is a narrative filled with triumphs and tragedies, bold ambitions and harsh realities. From the early expeditions of the 19th century to the scientific missions of today, the pursuit of the North Pole reflects our enduring fascination with the extremes of our planet and our desire to push the boundaries of knowledge and achievement.

The Triumphs and Tragedies of Arctic Exploration

The history of Arctic exploration is a mosaic of human triumphs and tragedies, etched against a backdrop of ice and endless night. It is a narrative that spans centuries, driven by the quest for knowledge, the allure of uncharted territories, and the relentless spirit of adventure. This chapter delves into the remarkable stories of those who dared to venture into the Arctic, celebrating their successes and mourning their losses.

In the early days of Arctic exploration, the primary goal was to find the elusive Northwest Passage, a sea route connecting the Atlantic and Pacific Oceans through the Arctic Ocean. The passage promised a shorter route for trade between Europe and Asia, but it was hidden beneath a treacherous expanse of ice. British explorers were among the first to embark on this perilous quest. Martin Frobisher, an English seaman, made three voyages in the late 16th century, but his efforts were ultimately in vain. He returned with stories of harsh conditions and ice-choked waters, yet his voyages paved the way for future explorers.

One of the most tragic and well-known Arctic expeditions was led by Sir John Franklin in 1845. Franklin, a seasoned explorer, set out with two ships, HMS Erebus and HMS Terror, and a crew of 129 men. Their mission was to navigate the final, uncharted sections of the Northwest Passage. Equipped with the latest technology and provisions for several years, the expedition seemed poised for success. However, the ships became trapped in the ice, and Franklin and his men were never seen alive again. The fate of Franklin's expedition remained a mystery for over a century, with search parties uncovering only scattered clues. It wasn't until the 21st century that the wrecks of Erebus and Terror were discovered, providing some answers but also raising new questions about the crew's final days.

The race to reach the North Pole itself became a central focus of Arctic exploration in the late 19th and early 20th centuries. This competition was marked by both triumph and controversy. American explorer Robert E. Peary claimed to have reached the North Pole on April 6, 1909, accompanied by his assistant Matthew Henson and four Inuit guides. Peary's claim was initially accepted, but later scrutiny and discrepancies in his records cast doubt on his achievement. Frederick A. Cook, another American explorer, also claimed to have reached the Pole a year earlier, in 1908. Cook's claim was widely disputed and lacked corroborative evidence, leading to a heated rivalry between the two explorers.

Despite these controversies, the quest for the North Pole continued. Norwegian explorer Roald Amundsen, known for his successful expedition to the South Pole, turned his attention northward. In 1926, Amundsen, along with Italian engineer Umberto Nobile and American aviator Lincoln Ellsworth, flew over the North Pole in the airship Norge. This marked the first verified flight over the Pole, showcasing the potential of air travel in Arctic exploration and providing a new perspective on the vast, icy expanse.

The mid-20th century brought significant advancements in technology, which in turn influenced Arctic exploration. In 1958, the United States submarine USS Nautilus made history by becoming the first vessel to reach the North Pole, traveling beneath the ice cap. This underwater journey highlighted the strategic importance of the Arctic during the Cold War and demonstrated new possibilities for exploration in one of the world's most inhospitable environments.

Another notable achievement came in 1968 when American explorer Ralph Plaisted and his team reached the North Pole by surface travel, using snowmobiles. Their journey, verified by satellite navigation, was a remarkable feat of endurance and innovation. Plaisted's expedition proved that modern transportation methods could conquer the Arctic's harsh conditions, opening the door for future surface expeditions.

The latter half of the 20th century saw a shift towards scientific exploration in the Arctic. The International Geophysical Year (1957-1958) and subsequent multinational collaborations aimed to study the Arctic's unique climate, geology, and ecosystems. These scientific endeavors underscored the importance of the Arctic in understanding global climate patterns and environmental changes. Researchers braved extreme conditions to gather data, contributing to our knowledge of this critical region.

However, the triumphs of Arctic exploration have often been accompanied by profound tragedies. The harsh environment, unpredictable weather, and shifting ice make the Arctic an unforgiving place. Many expeditions ended in disaster, with explorers succumbing to the cold, starvation, or accidents. The story of the Karluk, a ship that embarked on the Canadian Arctic Expedition in 1913, is a poignant example. Trapped in the ice, the Karluk drifted for months before being crushed. The crew faced a desperate struggle for survival, with only a handful managing to reach safety.

The 21st century has continued to see both triumphs and tragedies in Arctic exploration. Modern technology and improved safety measures have enabled more successful expeditions, but the Arctic remains a challenging and dangerous frontier. Climate change has added a new dimension to Arctic exploration, as melting ice opens new routes and reveals previously inaccessible areas. Scientists and explorers now face the dual challenges of studying the region's changing environment while navigating its still-treacherous conditions.

As we reflect on the history of Arctic exploration, it is clear that each triumph has been hard-won, and each tragedy has left an indelible mark. The explorers who ventured into the unknown did so with a mix of ambition, curiosity, and courage. Their stories of success inspire us, while their stories of failure remind us of the Arctic's unforgiving nature.

The legacy of Arctic exploration is not just in the maps drawn or the scientific data collected; it is in the human spirit's resilience and determination. The men

and women who braved the Arctic's icy expanses
pushed the boundaries of what was known and what
was possible. Their journeys, whether ending in
triumph or tragedy, have expanded our understanding
of the world and our place within it.

In a world where much of the unknown has been
mapped and explored, the Arctic remains a place of
mystery and challenge. The triumphs and tragedies of
those who came before us serve as both a guide and a
warning. They remind us that exploration is not just
about reaching a destination but about the journey
itself and the lessons learned along the way.

Roald Amundsen and the South Pole

Roald Amundsen's journey to the South Pole stands as
one of the most iconic feats in the annals of
exploration. Born in 1872 in Borge, Norway,
Amundsen was destined for a life of adventure. From
a young age, he was captivated by the tales of Arctic
explorers and the vast, icy expanses that beckoned
them. This fascination fueled his ambition to carve his
name into the history of polar exploration, ultimately
leading him to achieve what many had thought
impossible.

In 1910, Amundsen set his sights on Antarctica,
aiming to be the first to reach the South Pole. This was
a bold shift from his initial plan to explore the Arctic.
He kept his intentions secret, even from his own crew,
until they were well on their way. His decision was
partly influenced by the race against British explorer

Robert Falcon Scott, who was also vying for the same prize. This competition added a layer of urgency and determination to Amundsen's mission.

Amundsen's meticulous planning and understanding of polar conditions were key to his success. He chose the Fram, a ship designed for polar expeditions, which had proven its worth in previous Arctic explorations. The Fram's unique design allowed it to endure the crushing pressures of pack ice, making it an ideal vessel for the treacherous journey ahead. Amundsen also selected an experienced and resilient crew, essential for the grueling conditions they would face.

Upon reaching the Bay of Whales on the Ross Ice Shelf in January 1911, Amundsen established a base camp he named Framheim. From here, he and his team began the arduous task of preparing for their journey to the pole. Amundsen's strategy was methodical: he laid supply depots along the route, ensuring his team would have the necessary resources to sustain them through the harsh Antarctic environment. This careful preparation was a stark contrast to Scott's approach, which, while thorough, lacked the same level of logistical precision.

One of the critical decisions that set Amundsen apart was his choice of transportation. Understanding the efficiency of sled dogs in polar conditions, he brought a large number of them to pull the sledges. This decision proved crucial, as the dogs were not only resilient in the cold but also much faster than the ponies Scott used. Amundsen's intimate knowledge of dog-handling, gained from his earlier experiences in the Arctic, gave his team a significant advantage.

On October 19, 1911, Amundsen, along with four companions—Olav Bjaaland, Helmer Hanssen, Sverre Hassel, and Oscar Wisting—set out for the South Pole. They faced extreme cold, blizzards, and treacherous ice crevasses. The journey was grueling, testing their endurance and resolve. However, their meticulous preparations and use of sled dogs allowed them to make steady progress.

After a grueling trek of nearly two months, Amundsen and his team reached the South Pole on December 14, 1911. They planted the Norwegian flag, marking their triumph. The achievement was monumental, not just for its historical significance but for the demonstration of human perseverance and ingenuity in the face of nature's most formidable challenges. Amundsen's success was a testament to his leadership, strategic planning, and the unwavering spirit of his team.

Amundsen's return from the South Pole was equally challenging, yet his careful planning ensured they made it back to Framheim safely. News of his success reached the world in March 1912, and he was hailed as a hero. Meanwhile, Scott's expedition, which reached the pole a month after Amundsen, ended in tragedy. Scott and his team perished on the return journey, underscoring the perilous nature of Antarctic exploration.

Amundsen's triumph at the South Pole was a culmination of years of preparation and a deep understanding of polar environments. His ability to anticipate and mitigate the risks inherent in such an expedition set him apart from his contemporaries. He had learned valuable lessons from previous explorers, both in the Arctic and Antarctic, and applied this knowledge with precision.

Beyond the South Pole, Amundsen continued to push the boundaries of exploration. He was the first to navigate the Northwest Passage, and he later turned his attention to the Arctic, becoming the first to fly over the North Pole in the airship Norge in 1926. These achievements cemented his legacy as one of the greatest explorers of all time.

Amundsen's expeditions were not just about personal glory; they contributed significantly to our understanding of polar regions. His meticulous records and scientific observations provided valuable data for future researchers. His emphasis on preparation, teamwork, and respect for the harsh environments he explored set a standard for all subsequent expeditions.

Roald Amundsen's journey to the South Pole remains a landmark in the history of exploration. It is a story of vision, perseverance, and strategic brilliance. Amundsen's ability to adapt, his innovative use of technology and resources, and his leadership in the face of overwhelming odds continue to inspire modern explorers and adventurers. His legacy is a reminder of what can be achieved when human curiosity and determination are matched with careful planning and respect for the natural world.

As we look back on Amundsen's achievements, it is essential to recognize the contributions of his team and the indigenous knowledge he incorporated into his strategies. The success of the South Pole expedition was not the work of one man alone but a collective effort that highlighted the importance of collaboration and shared expertise.

Amundsen's story is a testament to the enduring human spirit and the relentless pursuit of knowledge and discovery. It is a narrative that transcends time, reminding us of the incredible potential within us all to overcome obstacles and achieve greatness. The lessons from Amundsen's expeditions continue to resonate, offering guidance and inspiration for those who dare to explore the unknown.

The Heroic Age of Antarctic Exploration

The Heroic Age of Antarctic Exploration, spanning from the late 19th century to the early 20th century, represents a period of relentless pursuit and extraordinary discovery. This era is defined by the courage and determination of explorers who ventured into one of the most inhospitable environments on Earth. These expeditions, driven by scientific curiosity, national pride, and the sheer desire for adventure, laid the groundwork for our understanding of Antarctica today.

The dawn of this age is often marked by the Belgica expedition of 1897-1899, led by Adrien de Gerlache. This Belgian expedition was notable for several reasons, not least of which was its multinational crew, which included the American physician Frederick Cook and the Norwegian explorer Roald Amundsen. The Belgica became trapped in the sea ice of the Bellingshausen Sea, forcing the crew to endure the harsh Antarctic winter. This unplanned overwintering provided valuable insights into survival in extreme conditions and marked the first time humans had spent a winter in Antarctica.

Ernest Shackleton, one of the most iconic figures of this era, embarked on multiple Antarctic expeditions. His first major venture was the British National Antarctic Expedition of 1901-1904, led by Robert Falcon Scott. This expedition, aboard the ship Discovery, aimed to conduct scientific research and explore the uncharted regions of the continent. Although Shackleton had to leave the expedition early due to health issues, the venture made significant contributions, including the discovery of the Polar Plateau.

Shackleton's most famous expedition was the Imperial Trans-Antarctic Expedition of 1914-1917. His plan was to make the first land crossing of Antarctica, but disaster struck when his ship, the Endurance, became trapped and was eventually crushed by pack ice. Shackleton's leadership during this ordeal became legendary. He kept his crew motivated and organized a daring open-boat journey to South Georgia Island to seek rescue. Remarkably, every member of his crew survived, a testament to Shackleton's extraordinary leadership and unyielding spirit.

Robert Falcon Scott, another pivotal figure, led the British Antarctic Expedition of 1910-1913, also known as the Terra Nova Expedition. Scott's goal was to reach the South Pole and secure it for the British Empire. The expedition faced numerous challenges, including extreme weather, difficult terrain, and logistical issues. Scott and his team reached the South

Pole on January 17, 1912, only to find that Amundsen had beaten them by just over a month. Tragically, Scott and his four companions perished on the return journey, a fate that underscored the perilous nature of Antarctic exploration.

Roald Amundsen's successful Norwegian Antarctic Expedition of 1910-1912 was a testament to meticulous planning and effective use of resources. Amundsen's approach to polar travel, which included the use of sled dogs and carefully placed supply depots, set a new standard for expeditions. His team's achievement in being the first to reach the South Pole on December 14, 1911, remains one of the greatest triumphs in the history of exploration.

The scientific contributions of these expeditions were immense. The Heroic Age saw the first comprehensive mapping of the Antarctic coastline, the discovery of new species, and significant geological and meteorological observations. These efforts laid the foundation for modern Antarctic science and provided valuable data that continues to inform our understanding of the continent.

The human stories from this era are equally compelling. The explorers faced unimaginable hardships: freezing temperatures, blizzards, isolation, and the constant threat of starvation. Their resilience and camaraderie in the face of such adversity are inspiring. The camaraderie among members, the leadership of figures like Shackleton, and the relentless pursuit of their goals despite the odds, are narratives that resonate deeply.

The expeditions of the Heroic Age were marked by a spirit of international cooperation and competition. National pride drove many of these ventures, but there was also a remarkable exchange of knowledge and techniques among explorers from different countries. This era highlighted the importance of collaboration in scientific discovery and exploration.

Technological advancements played a crucial role in these expeditions. The development of specialized ships, such as the Fram and the Endurance, capable of withstanding the harsh conditions of polar ice, was a significant achievement. Innovations in clothing, navigation, and survival techniques also contributed to the success and safety of the explorers. These technological strides not only facilitated the expeditions but also advanced our overall capabilities in extreme environments.

The legacy of the Heroic Age of Antarctic Exploration is profound. It set the stage for the Antarctic Treaty System, which preserves the continent for peaceful and scientific purposes. The international cooperation seen during this period foreshadowed the collaborative spirit that defines Antarctic research today. The scientific data collected during these expeditions continues to be a valuable resource for researchers studying climate change, glaciology, and marine biology.

One cannot overlook the personal sacrifices and the physical and mental toll on the explorers. Many faced frostbite, scurvy, and other debilitating conditions. The psychological strain of isolation and the constant battle against the elements required immense mental fortitude. These experiences highlighted the need for better preparation and support systems for future explorers.

The Heroic Age also left an indelible mark on popular culture and literature. The tales of bravery, endurance, and survival captured the public imagination and inspired countless books, films, and documentaries. The stories of Shackleton, Scott, and Amundsen have become part of the cultural fabric, symbolizing the human spirit's unyielding quest for discovery and understanding.

Modern Polar Expeditions

Modern polar expeditions have evolved dramatically from the days of the Heroic Age of Antarctic Exploration. Today, these ventures are characterized by advanced technology, international collaboration, and a strong emphasis on scientific research. Despite these advancements, the spirit of adventure and discovery remains intact, driving explorers and scientists to push the boundaries of human knowledge in the most extreme environments on Earth.

The logistical complexities of modern polar expeditions are immense, requiring meticulous planning and coordination. Unlike early explorers who relied on rudimentary maps and basic equipment, today's expeditions benefit from satellite imagery, GPS technology, and sophisticated communication systems. These tools allow for precise navigation and real-time data sharing, significantly reducing the risks associated with polar travel.

One of the most significant advancements in modern polar expeditions is the use of specialized vehicles and equipment. Icebreakers, such as the Russian vessel Akademik Fedorov or the Swedish Oden, are essential for navigating the thick sea ice that surrounds Antarctica and the Arctic. These ships are equipped with state-of-the-art laboratories, enabling scientists to conduct research while en route to their destinations. On land, tracked vehicles like the PistenBully and the Tucker Sno-Cat are employed to traverse the icy terrain, transporting personnel and equipment to remote research sites.

The role of aircraft in modern polar expeditions cannot be overstated. Planes and helicopters are used for reconnaissance, transportation, and emergency evacuations, providing a level of flexibility that was unimaginable to early explorers. The Antarctic continent, with its vast and inaccessible areas, particularly benefits from the use of aircraft. Planes like the Twin Otter and the Basler BT-67, specially modified to withstand extreme cold and high altitudes, are commonly used to ferry scientists and supplies between research stations.

International collaboration is a hallmark of modern polar expeditions. The Antarctic Treaty, signed in 1959, established Antarctica as a scientific preserve and banned military activity on the continent. This treaty, along with related agreements, fosters cooperation among nations, allowing for shared use of research stations and resources. The Arctic Council, formed in 1996, serves a similar purpose for the Arctic region, promoting sustainable development and environmental protection.

Scientific research is the primary focus of contemporary polar expeditions. Climate change, glaciology, biology, and astronomy are just a few of the fields that benefit from studies conducted in polar regions. The unique conditions of the poles provide insights into global processes, from atmospheric circulation patterns to the behavior of marine ecosystems. For instance, the European Project for Ice Coring in Antarctica (EPICA) has drilled ice cores that provide a record of climate changes over the past 800,000 years, offering invaluable data for understanding current climate trends.

The impact of climate change on polar regions has become a critical area of study. The Arctic, in particular, is warming at more than twice the global average rate, leading to significant changes in ice cover, permafrost, and ecosystems. Modern expeditions monitor these changes using a variety of methods, including remote sensing, automated weather stations, and underwater drones. The data collected helps scientists predict future changes and develop strategies for mitigating the effects of global warming.

The human aspect of modern polar expeditions has also seen significant advancements. The welfare of expedition members is a top priority, with modern clothing and gear designed to provide maximum protection against the extreme cold. Advances in nutrition and medical care ensure that participants can maintain their health throughout the often grueling conditions. Psychological support is also crucial, as the isolation and harsh environment can take a toll on mental well-being. Teams are trained to recognize and address issues such as depression and anxiety, ensuring that all members can contribute effectively to the mission.

Training for polar expeditions is rigorous and comprehensive. Participants undergo extensive preparation, including survival training, technical skills development, and team-building exercises. This training is essential for ensuring that all members are equipped to handle the challenges they will face. Simulated environments, such as cold chambers and altitude training facilities, help acclimate individuals to the conditions they will encounter in the field.

Modern polar expeditions also emphasize sustainability and environmental protection. The pristine environments of the polar regions are incredibly fragile, and measures are taken to minimize the ecological footprint of human activities. Waste management protocols, strict guidelines for scientific sampling, and efforts to reduce carbon emissions are all part of the commitment to preserving these unique ecosystems. Research stations are designed with sustainability in mind, incorporating renewable energy sources and efficient waste disposal systems.

Public engagement and education have become integral components of modern polar expeditions. Scientists and explorers share their experiences and findings through various media, including social media, documentaries, and public lectures. This outreach helps raise awareness about the importance of polar research and its implications for global environmental health. Educational programs and partnerships with schools enable students to learn about polar science and the challenges of exploration firsthand.

The technological advancements in data collection and analysis have revolutionized polar research. Instruments like autonomous underwater vehicles (AUVs) and remotely operated vehicles (ROVs) allow scientists to explore previously inaccessible areas beneath the ice. These tools provide detailed maps of the seafloor, capture high-resolution images, and collect samples from extreme depths. Satellite technology offers continuous monitoring of ice cover, weather patterns, and ocean currents, providing a comprehensive picture of the dynamic polar environment.

The role of women in polar expeditions has grown significantly in recent years. Historically, polar exploration was dominated by men, but today, women play key roles in all aspects of polar research and logistics. Initiatives to promote gender diversity and inclusion have helped ensure that the contributions of women are recognized and valued. This shift not only enriches the scientific community but also provides diverse perspectives that enhance the overall quality of research.

One of the most ambitious modern polar expeditions is the MOSAiC (Multidisciplinary drifting Observatory for the Study of Arctic Climate) expedition. Led by the Alfred Wegener Institute, this international collaboration involved the icebreaker Polarstern spending a year drifting with the Arctic sea ice. The goal was to study the Arctic climate system in unprecedented detail, gathering data on the atmosphere, ice, ocean, and ecosystems. The MOSAiC expedition exemplifies the scale and complexity of contemporary polar research, bringing together scientists from around the world to address critical questions about our planet's future.

The challenges of modern polar expeditions are immense, but so are the rewards. The knowledge gained from studying these remote and extreme environments has far-reaching implications for our understanding of Earth's climate, ecosystems, and geology. As technology continues to advance and international collaboration strengthens, the potential for new discoveries grows. The spirit of exploration that drove the early pioneers remains alive and well, inspiring a new generation of scientists and adventurers to venture into the unknown.

9 798330 339389